How Spacecraft Fly

Graham Swinerd

How Spacecraft Fly

Spaceflight Without Formulae

COPERNICUS BOOKS
An Imprint of Springer Science+Business Media

In Association with
PRAXIS PUBLISHING LTD

Graham Swinerd
University of Southampton
Hampshire, UK

ISBN: 978-0-387-76571-6   e-ISBN: 978-0-387-76572-3
DOI: 10.1007/978-0-387-76572-3

Library of Congress Control Number: 2008931301

Published in the United States by Copernicus Books,
an imprint of Springer Science+Business Media.

Copernicus Books
Springer Science+Business Media
233 spring street
New York, NY 10013
www.springer.com

Printed on acid-free paper

9 8 7 6 5 4 3 2 1

springer.com

This book is dedicated to the memory of
John Robert Preston
(1952–2007)

Foreword

The late science fiction author Sir Arthur C. Clarke, in an article published in 1945 in *Wireless World*, suggested that it would be possible to build a global communications system by placing artificial satellites in a strategically located orbit—the so-called geostationary Earth orbit. From their vantage point high above the Earth, the satellites would be able to relay information from any place to any other place around the world. Just twenty years later, Clarke's tentative proposal became a reality with the launch of the world's first commercial communications satellite—Intelsat 1. Now, a further four decades on from Intelsat 1, many of us would find it difficult to adjust to a world without satellites. Our cars and mobile phones routinely come equipped with satnav, the weather forecasts that we watch on our satellite TVs display images taken from space, Earth observation satellites monitor the threat of global warming and science spacecraft, such as the Hubble Space Telescope, have transformed our view of the universe. There can be no doubt that communications satellites, and the plethora of other satellites in diverse orbits, have profoundly affected the way we live.

But how are these satellites—these marvels of technology—designed? In what orbits can they operate? Once they are in orbit, how can we control them? What hazards do they face? How do we get them into orbit in the first place? Moving beyond Earth orbiting satellites, how can we design and propel spacecraft to rendezvous with comets or land on other planets? And, perhaps most importantly of all, what is the future of manned space exploration?

In this highly readable and entertaining book, Graham Swinerd shares with us his immense and personal experience of the first half-century of the Space Age, to answer these and many other questions—and all without recourse to mathematics!

Stephen Webb
Portsmouth, England
April 2008

Preface

As I write this introduction, it just happens to be 50 years since the launch of the first spacecraft. This dawning of the Space Age occurred in October 1957, when the Soviet Union lofted a small satellite called *Sputnik 1* into orbit. Since then, space activity has become an integral part of our culture, and from the perspective of the 21st century it is hard to appreciate what a major technical achievement and political coup *Sputnik* was. Although it did very little in orbit, other than to announce its presence through the transmission of a simple radio message, it nevertheless galvanized the other superpower, the United States, into a vigorous space program that ultimately led to men walking on the moon in 1969—just 12 years later!

When I was a boy, growing up in the early 1950s, my interest in space was sparked by an elementary school teacher, Mrs. Christian, and her inspirational gift of teaching science to her young class. When I look back over a long career in space, both in industry and academia, I have come to realize that this ball began to roll in that early classroom. I have a lot to thank that teacher for, who planted a lifelong interest and enthusiasm in me. At that time, the Space Age was yet to begin. Nothing was in orbit around Earth—apart from the moon, of course—and the exploration of the solar system was yet to begin. The only source of information about the planets had been gathered by astronomers through telescopes, and the only images of planetary landscapes were those produced by the space artist's brush.

How different it is today. Since the heady days of *Sputnik*, all of the planets of the solar system have been visited by robotic interplanetary spacecraft, with the exception of far-distant Pluto. Even as I write, this omission is being rectified by the launch of the *New Horizons* spacecraft in January 2006, which is due to fly by Pluto and its companion moon Charon in 2015. Ironically, in August 2006, just a few months after the launch, a gathering of astronomers in Prague stripped Pluto of its status as a planet, although the scientific objectives of the mission will of course not be compromised by this intriguing decision. Longer-term studies of the planets have also been undertaken by sending spacecraft to orbit the planets Venus, Mars, Jupiter, and Saturn. These missions have been extraordinary and surprising, having discovered a rich variety of features beyond our scientific expectations and imagination. Small bodies in the solar system, such as asteroids and comets, have also been the focus of recent space missions. One such example is a spacecraft called *Rosetta*, which was launched by the European Space Agency in March 2004 to rendezvous with, and orbit, a comet in 2014. As a consequence of all this activity, it is possible for the imagination and enthusiasm of today's schoolchildren to be stimulated by real photographic imagery from far-flung regions of the solar system.

Another space enterprise that has revolutionized our understanding of the universe is the launch of large space observatories into Earth orbit, where a clearer view of the cosmos is possible above the obscuring window of Earth's atmosphere. The most well known example of such a spacecraft is the *Hubble Space Telescope* (HST), which has revolutionized observational cosmology, to say nothing of the aesthetic quality of many of the images returned by the spacecraft. At the time of this writing, the lifetime of the HST is almost up, and the development of a second generation of large space telescope is currently underway. The new observatory, named the *James Webb Space Telescope*, will be launched around 2013 and have optics nearly three times larger than Hubble. It is hoped that the new telescope will be able to see the first stars and galaxies that formed after the Big Bang!

As well as all these scientific projects going on, there are a multitude of satellites in Earth orbit providing services to underpin the technological society we have here on the ground. These application satellites have become fully integrated into our lives, but without us really noticing that they are there. Perhaps the best example of this is global communications. If you talk with someone on another continent, your voice is most likely carried by a spacecraft in high orbit. Another example is satellite navigation ("satnav"), the use of which is rapidly spreading into business and leisure activities. At least in this case, we know that satnav has something to do with satellites. The other major application is Earth observation; there is an armada of spacecraft in low Earth orbit with imaging cameras, and other instruments, directed down to Earth's surface. Data from such spacecraft are used for everything from town planning to agriculture, and it is these satellites that give us a grandstand global view of things like climate change.

The final strand in all this is the presence of humans in space, which, apart from the Apollo astronauts reaching the moon in the late 1960s, has been confined to Earth orbit. Indeed, current activity is focused on the development of the *International Space Station* (ISS) in Earth orbit. When completed around 2010, the ISS will be the largest space structure ever built, weighing about 450 metric tonnes. However, many people look back at the

Apollo era and regard that as the golden age of spaceflight. As a consequence, the young people of today who are embarking on their careers not only have missed the main event, but also have not had the benefit of the inspiration that the Apollo era provided people of my generation. The moon landings were going on when I was in high school, and I have to say that Apollo was another reason (along with Mrs. Christian) why I chose to pursue a career in the space sector. Having said all that, it does seem that we are on the threshold of a new beginning for human space exploration. The planned retirement of the U.S. space shuttle fleet around 2010 is forcing a rethinking of American priorities in space, leading to the development of a program to return to the moon, and go on to land people on Mars within the next 30 years. This activity is also spurred on by the declared intention of other nations to return to the moon before 2020.

This book is a distillation of the knowledge and experience that I have acquired over my 30-year career in space. My main motivation is to share my enthusiasm with general readers, not just readers with a technical education. I have attempted to discuss all aspects of how spacecraft work, but in a way that is accessible to people who have an active interest in space but who do not have the scientific and mathematical background to understand the plethora of technical books that are available on this topic. I hope this book satisfies that interest and helps readers learn more about this truly fascinating subject.

The book discusses orbits, orbital motion, and weightlessness; how spacecraft are designed and how they work; and the likely developments in spaceflight in the 21st century, as well as a more speculative glimpse into the longer-term future of interstellar travel.

The book requires no prior knowledge on the part of the reader. There are no mathematical equations, and I have tried to explain everything in an understandable and physically intuitive way, although in a few cases I have had to simplify and generalize for the sake of clarity. The average reader with a nontechnical background will find the text comprehensible, challenging, and, I hope, enjoyable.

The idea of writing a book on spaceflight without resorting to mathematical equations arose from my involvement in short course teaching at the University of Southampton in England. Alongside all the teaching, research, and administration that are a normal part of a university academic's job, I have also been very much involved in professional development courses. Essentially, these are short training courses on space systems engineering, typically lasting 5 days or so, which we offer to professional engineers and scientists. Over the last 20 years, the European Space Agency (ESA) has been a principal customer in this business, and it has been a privilege over this period to visit ESTEC (ESA's technical headquarters at Noordwijk, The Netherlands) on many occasions as a course organizer and a lecturer. Usually these courses are attended by ESA staff members with a strong technical background, but in about 1995 the training department at ESTEC requested a new type of training program: a space engineering course for nontechnical staff! This was a radical departure from our usual training activity. But the ESA wanted to train its nontechnical employees, such as lawyers, accountants, contracts staff, and secretaries, in the technical aspects of the business in which they are involved to increase their motivation and productivity—a very enlightened training strategy. Over the years, this course has become very popular with ESA staff, being offered at a number of ESA venues across Europe. For us, the trainers, it posed significant challenges, in that we needed to put across to the delegates how spacecraft work without relying on prior technical knowledge or resorting to the use of mathematics. Meeting these challenges has been very rewarding, in terms of the appreciation of the course delegates, who have found a new fascination in learning how spacecraft fly. My wish is that readers of this book will find similar rewards.

This book is dedicated to John Preston, a dear friend who died in March 2007. Despite being very ill, John spent much time helping me by reviewing a partial manuscript of this book, which says so much about him as a person. Even John would have agreed that he was not a scientist. Having his view on the text, as someone steeped in the humanities, was particularly useful in helping me craft the text for people without a technical education.

One technical note: metric units of measure are followed by imperial units in parentheses, for the convenience of the reader. There is one exception: although a metric tonne differs from an imperial ton, which is the measure used in the United States (a metric tonne equals 1.102 imperial tons), the difference is slight, so the corresponding ton equivalent is not given.

Graham Swinerd
Southampton, England
October 2007

Acknowledgments

The writing of this book has taken longer than I anticipated, and there are many people who have helped directly and indirectly in the writing of it. I would like to thank:

John Preston and Stephen Webb for their reviews of the manuscript.

All the people at the publisher Praxis, but in particular Clive Horwood, for his continued encouragement and guidance for a rookie author.

My colleagues at the University at Southampton, especially the members of the Astronautics Research Group—Adrian Tatnall, Hugh Lewis, Guglielmo Aglietti, and Stephen Gabriel; of this group special thanks are due to Adrian, who has been a constant source of encouragement over my 20-year career at the university, and to Hugh, who has become a very supportive research partner in recent years.

Frank Danesy, who was Head of Recruitment and Training at ESTEC during the mid-1990s; the space engineering course for nontechnical staff at ESTEC, which gave rise to the idea for this book, was his brain-child.

Mrs. Christian, the elementary school teacher who set me off on the path of my lifelong love of "everything space."

My mother and my father (who died in 1995), who have supported me wholeheartedly at every twist and turn in my life's journey.

My children, Vicky and Jamie, for blessing and enriching my life beyond all measure.

Last, but not least, my wife, Marion, who has been by my side at every twist and turn, and who has always been my rock.

Contents

A Brief History of Space

A Primitive Model of the Universe

WE live in a world where it is difficult to extract ourselves from what might be called the collective knowledge of the human race. We are surrounded by huge information banks storing the collected thoughts of the clever people who have shaped the way we think about the world. Among these resources we can include our own education from our earliest years. In addition, we have the written word, the spoken word through television and radio, and access to the World Wide Web, which provides a view of the world through cyberspace.

As a consequence of this immersion in collective knowledge, most of us have an idea of the structure and workings of the solar system, and the universe in general, without ever having to venture beyond our armchair. It is very difficult, therefore, to put ourselves in the position of ancient man, who looked up at the sky with unsophisticated and unaided eyes, and attempted to make sense of it. If you could take this leap backward a few thousand years, and at the same time leave behind your collective knowledge, what would you make of it all? I would guess that only a minority of your friends and family would be interested anyway. Times were hard, and most of us would probably have spent the majority of our time just surviving. However, let's suppose that you were able to retain a few brain cells for things other than where your next meal was coming from. Perhaps the first obvious feature of your model of the universe would be a clear belief that the world was flat. If you look out of the window now, it does look flat, doesn't it? You wouldn't have needed to be too clever to notice that the Sun and Moon make a daily journey across the sky, from horizon to horizon in about 12 hours. If you were also a bit of an ancient astronomer, interested in the night sky, you also might have noticed that this daily journey is shared by the stars. You would have to be really observant, however, to have noticed that some of the brighter stars—what we now call the planets—wander among the fixed stars over a longer time scale.

What sort of model would you have dreamed up, as an ancient person,

G. Swinerd, *How Spacecraft Fly: Spaceflight Without Formulae,*
DOI: 10.1007/978-0-387-76572-3_1, © Praxis Publishing, Ltd. 2008

deprived of your collective knowledge? A sensible picture would be what might be called the "primitive" model—that the universe of Sun, Moon, and stars rotated about the flat Earth once every day. This flat Earth–centered model was indeed the accepted one for a long time, simply because it seems the obvious interpretation of what we see around us.

How we have moved on from this model to our current understanding of how the universe works is a well-documented but long and winding pathway through the history of science. Periodically, over time, a gifted individual has joined the journey along the pathway to challenge the accepted view. It is not the intention of this chapter to give a detailed account of this journey, but rather to look at some of the more important milestones, and to discuss some of the individuals who have made important contributions to the story.

Flat Earth to Spherical Earth

As we can see from the above, interestingly, ancient and modern civilizations can have widely different interpretations of the way things work, even though they both see the same sky. Generally, the differences occur as a consequence of the precision of observation that can be achieved using the unaided eye thousands of years ago compared to using powerful telescopes today.

The first substantive attack on our primitive model is credited to Eratosthenes, who lived in the ancient Egyptian city of Alexandria around 300 B.C. He used an astonishingly simple method, developed by a sharp intellect, to estimate the size of the spherical Earth, having first disregarded the belief that the Earth was flat. The basics of the method are illustrated in Figure 1.1a. Eratosthenes was somehow aware that at around the time of midsummer, vertical posts did not cast shadows at noon in Syene (point A), which is in a region of southern Egypt traversed by what we now call the Tropic of Cancer. However, at the same time of year and day, he could see that vertical posts in Alexandria (point B) did cast shadows. This supported the idea in his mind that the Earth was not flat, but spherical—an extraordinary leap of logic. He was also aware, using the geometry shown in Figure 1.1a, that the simple measurements of the angle a, and the distance between Alexandria and Syene, would allow him to measure the circumference of his now spherical world. The distance measurement was simple in principle, but arduous in practice, as he had to employ someone to pace out the 800 km (500 miles) or so between the two centers! His estimate of Earth's circumference was around 40,000 km (25,000 miles), which is amazingly close to our modern estimate of 40,075 km (24,903 miles).

Eratosthenes is remembered today for his ingenuity and vision, but also because he was right. It does make you wonder, though, how many of his fellow Alexandrians believed in his claims of a spherical Earth—something a bit hard to swallow for the average man in the street at that time. In order to draw his conclusion, he needed to assume not only that Earth was spherical, but also that the Sun was a long way away from Earth so that the sunlight illuminated Earth's surface with effectively parallel rays (Fig. 1.1a). An equally good interpretation of his observation of shadow lengths at noon is illustrated in Figure 1.1b, which would probably have gone down better with

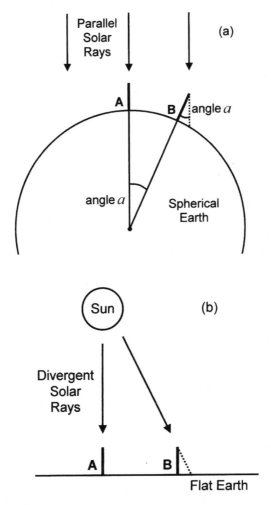

Figure 1.1: Alternative interpretations of Eratosthenes's observations of shadows cast by vertical posts at widely separated locations.

his contemporaries. If the Sun is assumed to be closer to Earth, so that the divergence of its rays is apparent, then the flat-Earth model can be saved. Over time, however, Eratosthenes was shown to be right, and the first cornerstone of our primitive model of the universe crumbled.

Earth-Centered to Sun-Centered Universe

The idea of an Earth-centered universe was firmly established by Claudius Ptolemaeus, known more commonly as Ptolemy, in the second century A.D. Ptolemy's universe had Eratosthenes's spherical Earth at its center, around which moved the Sun, Moon, planets, and stars. It was inconceivable that man, God's favored creation, should live anywhere other than at the center of the cosmos! Furthermore, by similar reasoning, it was supposed that these heavenly bodies, far removed from the imperfections of earthly life, should move along perfect circular paths.

However, there were problems with the model, which even the astronomers in Ptolemy's time could detect with their limited observational capabilities. The planets had been discovered centuries before—the Romans worshiped them as gods—and they could be distinguished not only by their brightness, but also by their movement across the sky relative to the fixed stars. Mars in particular appeared to challenge Ptolemy's model by moving erratically, performing loops in its motion among the stars, as shown in Figure 1.2. Ptolemy struggled to explain this behavior by introducing epicycles into his model. An *epicycle* is essentially a smaller circle around which a planet moves, which in turn is superimposed upon the larger circle representing the planet's motion about Earth. Throughout his lifetime, Ptolemy continued to tweak his model, introducing many epicycles in an attempt to fit observations.

Despite its evident weaknesses, the Earth-centered model survived for 1300 years or so, primarily because of the power and influence of the Church over this period. To challenge the notion that Earth was the center of the universe would have been considered foolhardy, a crime against God that could attract the severest penalty.

The person credited with making this challenge was Nicolaus Copernicus, a Polish Catholic cleric who was born in 1473. The main feature of Copernicus's universe was that he relegated Earth to be just one of a number of planets orbiting the Sun. At the time, this Sun-centered model was an extraordinary shift in our worldview, but Copernicus boldly swept away the old ideas, writing explicitly about the inadequacy of the previous arguments and refuting them. Copernicus waited until the year of his death, 1543,

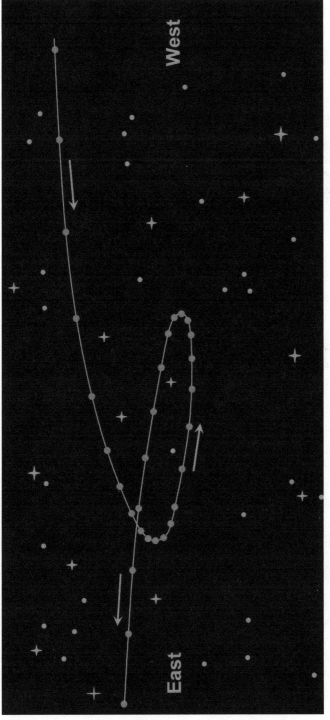

Figure 1.2: The apparent looping motion of Mars, relative to the fixed stars, as seen over a period of a few weeks.

before going public, presumably to avoid the consequences of religious persecution. Unkind contemporaries of Copernicus labeled him the "restorer" of the Sun-centered universe, in deference to Aristarchus, who held this belief around 280 B.C. However, the world was not ready for this idea in the third century B.C. Copernicus is remembered not just for establishing the idea of a Sun-centered solar system; many other related contributions secure his place in history:

- An understanding of the rising and setting of the heavenly bodies in terms of the daily rotation of the Earth.
- An explanation of the seasons due to Earth's annual journey around the Sun. Copernicus deduced that Earth's spin axis was not perpendicular to the orbit plane. Consequently, the Northern Hemisphere would be tilted toward the Sun during the Northern Hemisphere's summer, and conversely tilted away during the winter months.
- A mechanism to explain the looping motion of the planets among the fixed stars (Fig. 1.3).
- The estimation of the size of the planets' orbits in "astronomical units," and their periods (that is, the time taken to orbit the Sun). In this process, Copernicus assumed that the orbits were circular.

The last item on this list was a staggering achievement, and deserves further attention. First of all, what is an astronomical unit (AU)? In modern terms, it is the average distance between Earth and the Sun, taking into account that the distance varies a bit as the Earth orbits the Sun. Numerically

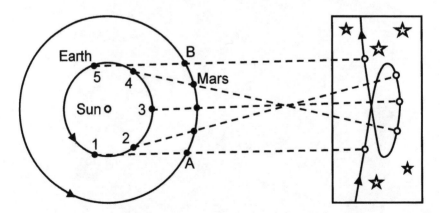

Figure 1.3: Copernicus's explanation of the apparent looping motion of Mars among the fixed stars. He assumed that Earth and Mars moved along circular orbits with different periods, so that Earth moved from point 1 to point 5 in the same time that Mars moved from point A to point B.

1 AU is around 150 million km (93 million miles). Copernicus had no way of determining this, but with careful thought he could devise ways of estimating the distance of the known planets from the Sun as multiples of the Earth-Sun distance—that is, in astronomical units. Therefore, he was able to construct the scale of the known solar system relative to the size of Earth's orbit, but its absolute size escaped him.

The explanation of his methods is a little complicated, but I hope the reader will come along for the ride! For the planets Mercury and Venus, closer to the Sun than Earth, the basics of this method are illustrated in Figure 1.4. Taking Venus as an example, Copernicus could observe its orbital motion around the Sun, as we can today, by watching its track in the sky at the time

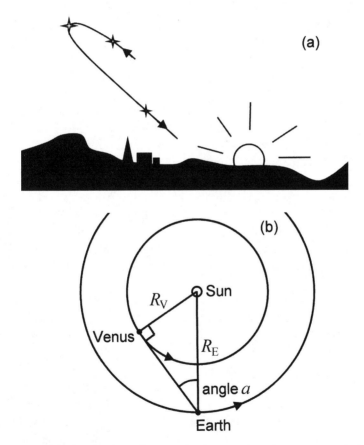

Figure 1.4: (a) The motion of Venus in the evening sky over a period of weeks, allowing the measurement of the maximum angle (angle a) between the planet and the Sun. (b) The orbital geometry of Earth and Venus at the time of maximum angular separation.

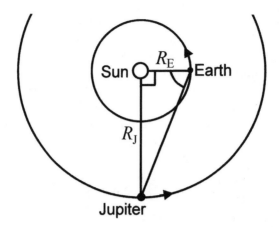

Figure 1.5: Copernicus estimated the radius R_J of Jupiter's orbit, again using 'simple' trigonometry.

of sunset over a number of weeks (Fig. 1.4a). To estimate the size of the orbit, Copernicus just needed to note the maximum angle between the Sun and the planet over this period, and he could then translate this to the orbital geometry shown in Figure 1.4b. The problem then reduces to a simple one, involving solving the lengths of the sides of a right-angled triangle using trigonometry. Most readers will have come across trigonometry in school mathematics lessons and probably will have forgotten it! However, referring to the triangle in Figure 1.4b, all that needs to be understood is that if Earth's orbit radius (R_E is 1 AU) and the maximum angle (angle a) are known, then the radius R_V of Venus's orbit can be calculated. Adopting this simple process, Copernicus found that R_V was approximately 0.7 AU, and a corresponding analysis of Mercury's motion gave its orbit radius as about 0.4 AU.

The process for estimating the orbit sizes of the outer planets known to Copernicus (Mars, Jupiter, and Saturn) was a little more involved. There are a number of ways of looking at this, but they all boil down to the same thing, and ultimately reduce again to a simple trigonometrical problem.Taking the planet Jupiter as an example, Copernicus measured the time it took for Earth to "lap" Jupiter in their respective orbits. He noted that approximately every 400 days Jupiter returned to the same position, due south in the sky at midnight. Translating this into orbital position, he realized that this happened when Earth was precisely between the Sun and Jupiter, and was about to overtake Jupiter. He then went on to deduce that a quarter of this lapping period—approximately 100 days—after this alignment, Earth would be 90 degrees ahead of Jupiter in its orbit, giving the orbital geometry shown in Figure 1.5. The measurement of the angle between the Sun and Jupiter at this time, an observation best made at sunset, completed the puzzle and allowed Copernicus to calculate the radius of Jupiter's orbit at about 5.2 AU. Similar calculations gave estimates of the radii of Mars's and Saturn's orbit, at around 1.5 AU and 9.5 AU, respectively.

When Johannes Met Tycho

Copernicus's work, containing a wealth of apparently irrefutable detail, put the Earth-centered universe to rest, and finally removed the constraints that had inhibited the quest to understand the solar system for over a millennium.

The next person to make progress on this quest was Johannes Kepler, who was born in Germany in 1571. As a theoretician of the first order, he brought his intellect to bear upon Copernicus's model of the solar system, and found it lacking. However, Kepler knew that precise observations of the planets' motions were required in order to expose the weaknesses of Copernicus's model and make further progress along the pathway. This need was satisfied by Kepler's chance association with Tycho Brahe, a Danish nobleman who spent much of his life and resources developing an astronomical observatory on an island off the coast of Denmark. This housed precision instruments, and Tycho compiled what was the most complete and accurate catalogue of planetary position measurements available at that time.

Johannes's brilliance as a theoretician and Tycho's observational genius complemented each other perfectly, to bring about the next revolution in understanding. However, their relationship was an uneasy one, and Tycho was reluctant to gift his life's work to a younger rival. Tycho did make his observations available to Kepler, but only in a frustratingly piecemeal manner. This impasse was finally resolved on the death of Tycho, after which Kepler was able to extract the full catalogue of measurements from Tycho's family.

Now that Kepler had accurate observations, he spent a number of years trying unsuccessfully to reconcile them with the notion that planetary orbits were circular. Looking at the orbit of Mars, he struggled for nearly a year to resolve a discrepancy between observation and theory of only 8 minutes of arc—a small angular measure of about one-quarter the diameter of the full moon. This in itself says a great deal about Kepler's integrity and honesty; clearly, it would have been easier to ignore such a small anomaly, or to regard it as an erroneous measurement. This struggle, however, led Kepler to the idea that was to be his core contribution to the understanding of the solar system—that planetary orbits were elliptical in shape. Making this step, he now found that Tycho's measurements fitted beautifully, and thereafter Kepler published his first two laws of planetary motion in 1609. His third law, to do with the relationship between the size of an orbit and its period, was also a tough one that took him a further 10 years to establish. Kepler's three laws of planetary motion are as follows (see also Figure 1.7):

Kepler 1 – The orbit of each planet is an ellipse, with the Sun at one focus.
Kepler 2 – The line joining each planet to the Sun sweeps out equal areas in equal times.
Kepler 3 – The square of the period of a planet is proportional to the cube of its mean distance from the Sun.

It is worth dwelling a few moments on Kepler's laws, to explain the jargon, and to illustrate their meaning. The first law uses the word *ellipse,* which from high school geometry could be described as egg-shaped or a squashed

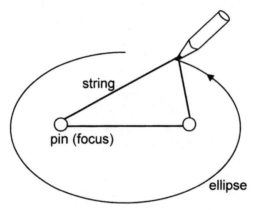

Figure 1.6: Drawing an ellipse. The ellipse's focal positions are where the pins penetrate the card.

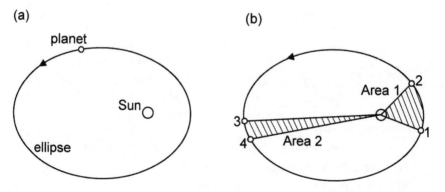

Figure 1.7: Illustrations of (a) Kepler's first law and (b) his second law.

circle. We know that to draw a circle, we use a compass. The resulting figure has one focus—the center where the point of the compass penetrates the paper. You may also have drawn an ellipse in school by pressing two pins into a piece of card, and placing a loose loop of string around the pins. Placing a pencil in the loop, and keeping the string tight, we can move the tip of the pencil over the paper to produce an ellipse, as shown in Figure 1.6. The two points where the pins penetrate are referred to as the focuses of the ellipse. From Kepler's first law (refer to Fig. 1.7a), we see that planets move along elliptical orbits, but also that the Sun is located at one of the focal positions.

Kepler's second law is a rather strange way of describing how fast a planet moves at different points on its orbit. Looking at Figure 1.7b, we can see that if a planet moves from point 1 to 2 in the same time that it takes to move from point 3 to 4, then Kepler's second law implies that the shaded areas— Area 1 and Area 2—must be equal to one another. This geometrical argument can be translated into a dynamical one, since it is easy to see that, for this to happen, the planet must move rapidly when close to the Sun, and more slowly when further away. Kepler's 17th century mind tended to think in terms of geometry, whereas a modern orbit analyst would tend to take the dynamical view.Kepler expressed his third law in terms of a word equation, and since we are trying to avoid the use of equations it is sufficient to say that planets in big orbits take longer to orbit the Sun than planets in small orbits—a fairly commonsensical notion.

Trying to summarize Kepler's activities in a few paragraphs, as we have done here, does no justice to the magnitude of his achievement in establishing the modern view of the way the planets move around the Sun. The time he took to do this does, perhaps, give a measure of the difficulty of the task. His achievement is emphasized by noting that his laws are still used today when engineers analyze the motion of spacecraft around the Sun, or indeed the orbit of a satellite around the Earth. It is also important to realize that Kepler developed his laws empirically, based purely on Tycho's catalogue of planetary measurements. He described *how* the planets moved around the Sun, but had no underlying theoretical foundation to explain *why* they moved in this way. This task was left to the intellectual giant that was Isaac Newton.

"On the Shoulders of Giants ..."

Newton was born on Christmas Day 1642, just 23 years after Kepler had published the last of his three laws—and what a gift to the world! Newton has been described in clichéd terms by numerous biographers: "the foremost scientific intellect of all time"; "the father of modern science"; and so on. However, when applied to Newton, these glowing epithets are arguably fully justified. Newton made major contributions to many areas of scientific activity, including optics and light, mathematics, dynamics, gravitation, and theoretical astronomy. He himself, however, summarized his contribution to science by stating, "If I have been able to see further, it was only because I stood on the shoulders of giants," referring to some of those giants discussed in the preceding sections of this chapter.

Newton's quest to understand the world began with his undergraduate career at Trinity College, Cambridge University, at the age of 18. However, his time at Cambridge was interrupted in the summer of 1665, when the university was closed down by an outbreak of plague. Newton then returned to his birthplace, the isolated village of Woolsthorpe in Lincolnshire, where in an amazingly productive period of 2 years he revolutionized science.

In summary, during this period he devised his law of gravitation and his laws of motion. Combining these, he was able to formulate the equations that governed the motion of the planets around the Sun. He then realized that these equations could not be solved using the methods then available. However, he was not a person to let such a small detail inhibit his efforts, and so he set about inventing a new branch of mathematics, called calculus, to remove the barrier. All of these accomplishments have had a lasting impact upon science and engineering to the present day, and any one of them would be considered a major intellectual achievement. For them all to have come from one individual in such a short period of time is extraordinary.

It is worth pausing a few moments to consider in more detail each of the steps that comprised Newton's achievement. Perhaps the thing most people associate with Newton is his law of gravitation, along with the story of the mythical apple that is supposed to have fallen on his head and given spontaneous birth to the idea. It is likely, however, that the formulation of his understanding of gravity took a little more time and effort. The formal statement of Newton's law of gravitation is given below, and as can be seen it is expressed once again as what might be called a word equation:

Newton's Law of Universal Gravitation – the force of gravity between two bodies is directly proportional to the product of their masses, and inversely proportional to the square of their distance apart.

However, it can be easily understood in simple terms. The phrase "directly proportional to the product of their masses" simply means that the force of gravity between two large objects—say two planets, or two stars—is large, and indeed will govern the way these celestial bodies move with respect to each other. On the other hand, the force of gravity between two small objects will be tiny. For example, if you place a couple of balls on a pool table, you expect them to remain firmly attached to the surface, since they are attracted to the large mass which is the Earth beneath the table. It is only the structural strength of the table that is preventing them from responding to the force by whizzing off toward Earth's center. At the same time, we do not expect them to move across the table toward each other, since the force of gravity between them is so small as to be effectively zero. The game of pool would be somewhat different if it were otherwise!

The way the force of gravity varies with distance, as described above, is sometimes referred to as the *inverse square law*. This describes how the force between two bodies diminishes as they move further apart. If you think of two objects a particular distance apart—strictly this distance is measured between their centers—then the force of gravity between them will have a particular strength. When we move them apart so that the distance between them is doubled, the inverse square law says that the force is one fourth of what it was before. To get this, we take 2 from "twice the distance," square it to give 4, and then take the inverse to give us the one fourth. In the same way, we can move the bodies 10 times further apart, and the same argument tells us that the force of gravity is reduced by a factor of $\frac{1}{100}$.

There is some debate among scientific historians about how Newton settled upon the inverse square law for gravitation. Some believe he was influenced by his studies of the way light behaved; he discovered by experiment that the intensity of light falling upon a surface decreased in proportion to the inverse square of the distance between the source of light and the surface. However, more likely he proposed the inverse square law since it was consistent with Kepler's third law of planetary motion, which can be shown by the use of some simple mathematics that can be done literally on the back of an envelope.

Coming back to Newton's apple, we can explore some of Newton's thinking during his brief but prolific period of exile in the Lincolnshire countryside. Having thought about his law of gravitation in a universal context, Newton's observation of the fall of an apple from a tree engendered universal questions in his mind such as, Why doesn't the moon also fall to the ground? To answer this one, we can compare the motion of the apple with that of the moon.

Taking the apple first, when it is released from the tree it responds to the force of gravity by accelerating toward the ground. It starts from rest up in the branches of the tree and builds up speed until impact with the ground. If we were able somehow to measure this impact speed and the time of fall, we would be able to calculate its acceleration. For example, if the height of the tree was such that the apple took 1 second to hit the ground, we would find that its impact speed was about 10 meters per second (32 feet per second). In this case, the distance fallen can be estimated as about 5 meters (16 feet)— quite a tall apple tree. If the ground were not in the way, the apple would continue to accelerate toward Earth's center, gaining 10 meters per second in speed for every second of the fall. This acceleration due to gravity at Earth's surface of 10 meters per second per second is usually expressed as 10 m/sec/sec or 10 m/sec^2 (32 feet/sec^2).

Newton was the first to realize that the moon must also respond to the force of gravity in the same way. However, the moon is around 60 times more distant from Earth's center than his apple. Applying his law of gravitation, he estimated that the acceleration of the moon toward the Earth will be much less than that of the apple by a factor of $^1/_{3600}$—that is, the inverse of 60 squared. In its distant orbit then, the moon will fall toward Earth with an acceleration of approximately 10/3600 meters per second per second, or about 3 millimeters/sec^2 ($^1/_{100}$ feet/sec^2). With this small acceleration downward, it is easy to estimate that in 1 second the moon falls a small distance toward Earth of about 1.5 millimeters—much less than the 5 meters fallen by the apple. However, it is instructive to consider what happens to the moon's motion during the period of a minute, as then the numbers are a little easier to grasp. Because of the coincidence that the moon is 60 times further away from Earth's center than the apple, and that there are 60 seconds in a minute, the mathematics tell us that the moon falls the same distance in 1 minute as the apple falls in 1 second—about 5 m. However, at the same time the moon has a relatively high speed along its orbit so that in 60 seconds it moves horizontally approximately 61,100 meters (200,500 feet). Figure 1.8 shows that the combination of these horizontal and vertical motions result in the near-circular orbital path that we observe, so that although the moon does actually fall continually toward Earth, fortunately it never reaches the ground!

To understand the motion of bodies, such as the apple and the moon, in this way, Newton had to devise not only his universal law of gravitation, but also his three laws of motion, which are stated as follows:

Newton 1 – A body will continue in a state of rest, or of uniform speed in a straight line, unless compelled to change this state by forces acting upon it.

Newton 2 – The rate of change of momentum of a body is proportional to the force acting upon it, and is in the same direction as the force.

Newton 3 – To every action there is an equal and opposite reaction.

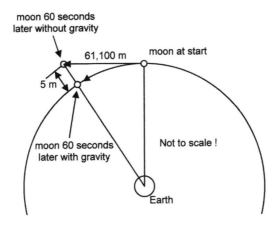

Figure 1.8: Newton applied his universal law of gravitation to the Moon, as well as to his apple, to show how the Moon orbits Earth.

It is important to note that these laws, written in the form of words here, have their most powerful manifestation when expressed in mathematics. They revolutionized 17th century science, and indeed still dominate engineering science today. Newton's contribution is summed up by noting that 21st century engineers still use the mathematical expression of these laws to design buildings, bridges, cars, airplanes, and indeed spacecraft. I would love to stand with Isaac Newton at the end of a modern airport runway as a jumbo jet is taking off, and tell him that he is responsible for this apparently impossible apparition of 350 metric tonnes of predominately metal soaring into the sky—an amazing legacy!

In terms of our understanding of the solar system, Newton's revolution came about when he combined his law of gravitation with his laws of motion to produce equations that described the motion of the planets around the Sun. As described earlier, the solutions of these equations were obtained only after Newton had devised a new branch of mathematics. But once this was done, Newton rediscovered Kepler's three laws of planetary motion in his mathematics, thus giving a theoretical basis to Kepler's empirical work completed almost half a century before. However, Newton found not only Kepler's work in his new formulism but lots more. His mathematics were saying that objects moving in a gravity field, for example, a planet moving around the Sun, or a spacecraft moving around a planet, were not confined to elliptical paths. The shape of the path of such an object could also be that of a circle, a parabola, or a hyperbola. Most people are familiar with circles and ellipses, but what about the parabolic and hyperbolic shapes? These four shapes are referred to as conic sections,

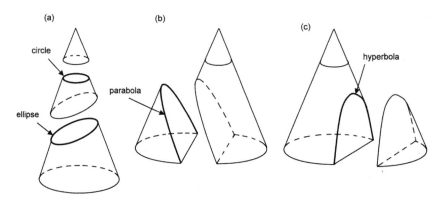

Figure 1.9: The shapes of orbital trajectories can be obtained by slicing a cone.

because you can get them by sectioning, or slicing, a cone as illustrated in Figures 1.9 and 1.10.

In Figure 1.9a, we can see that if we slice the cone horizontally we get a circle, and if we slice it slightly obliquely we get an ellipse. Another trajectory shape is obtained by slicing the cone in such a way that the plane of the slice is parallel to the side of the cone, as in Figure 1.9b. In this case we get a parabola. The parabolic trajectory is not a closed one—as are the circle and ellipse—so we have to imagine that the cone is infinitely big, and not truncated at the base as illustrated. An example of an object on a parabolic trajectory is a comet that falls with initially zero speed from infinity—or effectively from a great distance—and is swung around the Sun and ends up heading back to the same place, arriving again at infinity with zero speed. Perhaps this type of trajectory can best be describes as a celestial U-turn.

The final trajectory shape that Newton found in his mathematics is the hyperbola, which is obtained by slicing our cone vertically, as shown in Figure 1.9c. This trajectory is again an open one, stretching to infinity, so we have to imagine that our cone is very large. An example of this type of orbit is a spacecraft swinging by a planet. The spacecraft approaches from great distance, initially traveling at constant speed relative to the planet in a straight line, as the gravitational influence of the planet is tiny. However, as the spacecraft closes in, the gravitational force increases and the trajectory is deflected. The gravity of the planet swings the spacecraft's path around so that the vehicle leaves the planet in a new direction, traveling in a straight path again at the same planet-relative speed once it has reached great distance. As we will see later in Chapter 4, this type of swing-by trajectory is commonly used by engineers designing interplanetary spacecraft missions. The hyperbola is distinguished from the parabola by the deflection angle; in a parabolic trajectory the object is deflected by 180 degrees by its encounter

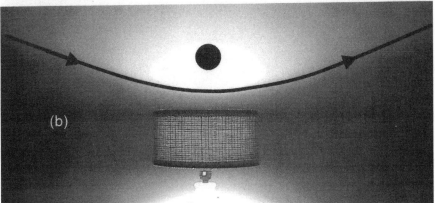

Figure 1.10: The light cast by a table lamp shows the hyperbola-shaped, swing-by trajectory of a spacecraft.

with the planet, whereas the hyperbolic trajectory is deflected by an amount less than this.

Surprisingly, the hyperbola is also a fairly common sight in everyday life. All you need is a lamp with a circular shade, which projects a cone of light both upward and downward, producing a circle of light on the ceiling above and on the table below. If, however, the lamp is placed next to a wall, then the cone of light is effectively sliced vertically, as in Figure 1.9c, to produce the shape of a hyperbolic trajectory on the wall. A photograph of such a shape is shown in Figure 1.10a, and how it relates to the orbital trajectory is shown in Figure 1.10b. If you find yourself on a dinner date in a restaurant with this common type of wall lighting, you could use your knowledge of celestial mechanics to break the ice. (On the other hand, your guest may just think that you are a rather sad person who needs to get out more!)

Newton himself thought of his scientific investigations as a small contribution toward revealing the fundamental laws of the universe that were "written" by God, its designer. Thinking about his discoveries in this context, it *is* rather strange that we live in a universe where the shapes of gravitational trajectories are related to slices of a cone!

The final episode in this story of Newton's achievement is also surprising. Having "solved the universe" in this way, Newton then failed to communicate his work to anyone! Meanwhile, unknown to him, contemporary scientists—principally Robert Hooke and Edmund Halley (of Halley's Comet fame)—were struggling with the problem of planetary motion in the coffee houses of London. Finally in 1684, Halley visited Newton in Cambridge, hoping to gain some insight into the riddle. When Halley posed the question about the shape of gravitational trajectories about the Sun, Newton revealed that he had already solved the problem, but had characteristically misplaced it. It was ultimately Halley who encouraged Newton to write his landmark work, the *Philosophiae Naturalis Principia Mathematica*, requiring 2 years of hard work to complete. In this rather circuitous manner, Newton was finally recognized as being one of the greatest scientific thinkers of all time.

The only other individual described is this way is perhaps Albert Einstein, whose scientific genius was unleashed upon the world at the beginning of the 20th century.

What Did Einstein Do for Us?

Einstein's contribution was fundamental and profound, a revolution in the way we think about the physics of motion, and in particular the motion of bodies in a gravitational field. This revolution began with the publication of Einstein's *special theory of relativity* in 1905, when Newtonian physics was well established, and most scientists believed their understanding of the physical laws of nature was complete. After all, newtonian physics had reigned supreme for something like 220 years! This blow to the scientific establishment was all the harder to take, as Einstein's interest in physics was a hobby at the time; his job was that of a patent clerk in an office in Berne. However, his new physics took the scientific community by storm.

A cornerstone of Einstein's work was an appreciation that the arena in which all physical events take place is a four-dimensional world called *space-time*. In other words, to describe the location of a physical event—for example, the impact of an apple on the ground—we need four numbers, three defining its position in space, and another giving the time. In Newton's physics the three-dimensional spatial world and time were considered to be independent and absolute. However, in Einstein's theory, space and time are inseparably interwoven, and the place and time defining an event are not absolute but depend on the state of motion of the observer. This rather strange notion led to the uncomfortable idea that Newton's physics was incorrect; however, the differences between Newton's and Einstein's descriptions of the world manifested themselves only when things moved at very high speeds, that is, speeds near the speed of light of 300,000 km per second (186,000 miles per second).

Einstein's revolution was not complete, however, as in 1916 he published his theory of gravitation—the *general theory of relativity*. The journey from the special theory to the general was not an easy one, and Einstein struggled with the physics and, in particular, the mathematics required to formulate his gravitational theory. Indeed, the mathematics required to describe his theory of gravity were so complex that it was claimed that few people in the world actually understood it when first published. Fortunately, the principles of the theory can be explained in relatively simple terms.

Einstein's description of the way planets moved around the Sun is completely different from Newton's view. In Einstein's theory, the four-dimension world of space-time is not just a background reference system against which the locations and timings of physical events are recorded, but rather it becomes a dynamic entity, playing a central role in the way things move in a gravity field. The underlying principle of Einstein's general theory is that massive objects, like the Sun, distort the geometry of space-time. This is the famous *warped space*, which has become so familiar to us all, courtesy of popular science-fiction epics like *Star Trek*. However, although we have heard a lot about it in sci-fi stories, nevertheless an appreciation of what a curved four-dimensional space-time continuum means is very difficult to grasp, even for those equipped to understand the mathematics! Einstein's basic idea of motion in a gravity field is that objects move in such a way as to take a path that gives the shortest distance between two points. Clearly in our everyday experience, the path defining the shortest distance between two points is a straight line. But then, in our everyday experience, we do not often come across warped space!

However, there is one everyday example of determining the shortest distance between two points in a curved space—that is, the efficient global routing of aircraft. For example, what is the shortest distance between London and Sydney in the curved two-dimensional space we call Earth's surface? If we take a map and just draw a straight line between London and Sydney (the broken line in Figure 1.11), we find that this is not the shortest route. The shortest route can be found by stretching a piece of string on a globe, holding down the ends over London and Sydney. If you then plot this route on a map (the continuous line in Figure 1.11), you will find that the shortest route is curved. A quick experiment with a globe and a piece of string will help you to check this.

Returning to Einstein's gravity, as we said above, the influence of the massive Sun is to produce curvature in the fabric of space-time surrounding it. Straight lines in this space are no longer straight, but curve along the contours of the warped space produced by the Sun. The resulting orbital trajectories are effectively those found by Kepler and Newton. Given this, the reader might ask why Einstein's complex theory of gravity is needed, when Newton does a perfectly good job already. The answer is that Einstein's theory goes further, and predicts additional effects that are particularly conspicuous in very intense gravitational fields. A good example of this is the bending of light as it passes the Sun, an experimentally confirmed effect that is not predicted at all by Newton's theory. In our everyday experience, a beam of light is perhaps the best way of defining a straight line. However, in

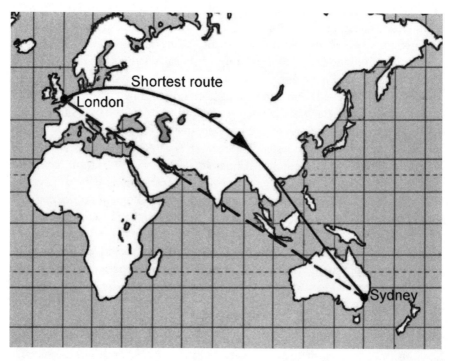

Figure 1.11: In the curved two-dimensional space of Earth's surface, the shortest distance is not a straight line.

the warped space surrounding the Sun, the path of the light is deflected (very slightly) in response to the curvature of space-time. Another curious feature of Einstein's general theory is that when space-time is curved by the presence of a massive object, not only are the spatial dimensions curved, but the time dimension is as well; we are presented with the bizarre notion that clocks run at different rates depending on how close they are to the object!

Clearly Einstein's achievements are pertinent to our story, but we should return to the question in the title of this section: What did Einstein do for us? Well, if the "us" refers to spacecraft design engineers, the honest answer is "not a lot!" The spacecraft missions achieved in the first half century or so of the Space Age involve space vehicles that have not achieved very high speeds, compared to the speed of light. Similarly our activities have effectively been confined to the region of space near the Sun, where very intense gravitational fields are not encountered. As a consequence, the more exotic effects of Einstein's theory do not manifest themselves, and we are left with the rather surprising conclusion that modern spacecraft engineers still use 300-year-old Newtonian theory.

There is, however, one clear example where Einstein's relativity theory does make an essential contribution to the design of a spacecraft. The U.S. Department of Defense operates a space system called Navstar Global Positioning System (GPS), which is used as a navigational aid for all branches of the U.S. armed forces. However, a lot of people reading this may have used GPS for leisure purposes—hiking, sailing, or flying—or for in-car navigation. The space system comprises a constellation of 24 satellites in near-circular orbits at heights of around 20,500 km (12,700 miles). If you have an appropriate receiver on the ground, the system will provide information about your location accurate to about 10 meters in each of the three spatial dimensions. To do this, however, each satellite must carry an atomic clock, which needs to be accurate—to about one second in every 30,000 years or so! To do the necessary calculations to find your position on the ground, your receiver must also have a clock. Fortunately, this clock need not be quite so sophisticated (or expensive!) as the satellite clocks, but it should record the passage of time at the same rate as the orbiting clock, during the short period when the receiver is doing its calculations to estimate where you are. However, the receiver clock on the ground is a lot closer to the gravitational mass of Earth than the satellite clock, and therefore Einstein said the ground clock will run slower than the orbiting clock. Over the period of a day, the combined effects of Einstein's theory cause an accumulated error of around 38 microseconds (38 millionths of a second) difference between the orbiting and ground clocks. Although this sounds small, when translated into a navigational error it amounts to about

10 km. After a day your in-car navigation system might be indicating that you are in the wrong town!

When the first experimental GPS satellite was launched, some engineers were skeptical about the importance of the Einstein effects, but soon realized that time warping is a reality. To overcome this problem in the current spacecraft design, the satellite clocks are manufactured with an appropriate offset in the clock rate built in.

How we have come to understand space is a rather intriguing story, and what I have presented here is an abbreviated and personal view of something that could have a whole book devoted to it. In summary, perhaps one of the most surprising conclusions to be drawn from this discussion is that modern spacecraft engineers still predominantly use Newtonian theory to design spacecraft, and to design the orbits they travel to achieve a particular destination. We will take this notion forward in subsequent chapters, where the way spacecraft are designed is discussed in more detail.

Basic Orbits

How Spacecraft Move in Orbit

BEFORE getting to the business of discussing the orbital aspects of modern spacecraft missions later in this chapter, there are a few fundamentals about the orbital motion of a spacecraft that we need to discuss, and a few popular misconceptions about it that need to be put to rest.

The first of these fundamentals is how a spacecraft remains in orbit around Earth, effectively forever, without having to fire rockets to sustain the motion. The answer lies in understanding that the spacecraft, like a stone falling down a deep well, is in a state of continual free-fall. Clearly, we do not expect the stone's motion to be assisted by rockets; it just falls unaided in the gravity field until it impacts the water at the bottom of the well. Free-fall in a gravity field is also the key to understanding the spacecraft's motion, although in this case it is perhaps not so apparent. And, of course, the spacecraft operator hopes that, in the process, it does not impact the ground like the stone!

To help with this discussion, we turn to a device that has become known as *Newton's cannon*, after its originator Isaac Newton. He first introduced the idea around 1680 in *A Treatise of the System of the World*, which he wrote as a popularization of his great work the *Principia* (see Chapter 1). Newton produced a diagram of his cannon in his treatise similar to that in Figure 2.1. To start with we have to imagine an impossibly high mountain, let's say 200 km (124 miles) high, for the sake of argument—a real challenge to the climbing fraternity. Not only is it a long way to the top, but when you get there you are effectively in the vacuum of space. Then you have to envisage dragging all the materials necessary to the summit to build a large cannon there that is capable of firing projectiles at a range of barrel speeds. This is also illustrated rather unimaginatively in Figure 2.1.

The cannon crew, presumably all dressed in space suits, now begins the serious business of firing cannonballs at the unsuspecting population below. You can see that if the crew fires a cannonball out of the gun with a barrel speed of, say, 2 km/sec (1.24 miles/sec), then it will do as you expect it to –

G. Swinerd, *How Spacecraft Fly: Spaceflight Without Formulae,*
DOI: 10.1007/978-0-387-76572-3_2, © Praxis Publishing, Ltd. 2008

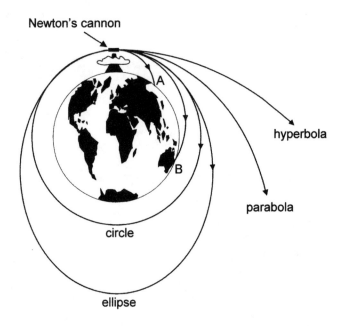

Figure 2.1: Newton's cannon—a "thought experiment" devised by Isaac Newton to explain the nature of orbital motion.

that is, take a curved path and impact the ground some distance away at point A (Figure 2.1). If the barrel speed is then ramped up to, say, 6 km/sec (3.73 miles/sec), the cannonball becomes intercontinental and travels a whole lot further to point B before impacting Earth's surface. However, something really interesting happens when the crew further increase the barrel speed to around 8 km/sec (5 miles/sec). Now the cannonball's path curves toward Earth's surface once again, but the curvature of the trajectory is matched by the curvature of Earth, and the ball continues to fall toward Earth without actually making contact! The projectile has now entered a circular orbit around Earth (Figure 2.1), and the cannon crew had better watch out as the ball will whizz past the gun about 90 minutes later, after making a complete orbit of Earth. Like the stone, the ball is now in a state of free-fall, and will continue to orbit Earth indefinitely.

To reinforce this idea, we can look at the same situation but consider Earth's curvature. At the mountain summit, the curvature is such that Earth's surface falls away below a truly flat horizontal plane by about 5 m (16 ft) in approximately every 8 km (5 miles) traveled over the ground. You may also recall that 5 m is roughly the distance fallen by Newton's apple in 1 second (Chapter 1). If you fire a cannonball from the gun at an initially horizontal speed of around 8 km/sec, it too will fall 5 m in the first second of flight, thus matching Earth's curvature. Therefore, no ground impact results, and you have orbital motion. To be more precise, for the cannonball to enter a circular orbit, it must be fired at a speed of 7.78 km/sec (4.84 miles/sec) from the summit cannon. For those of you who like more familiar units, this is about 28,000 km per hour (17,400 mph), which is a typical speed for a space shuttle in low orbit.

Newton's other orbital trajectories (Chapter 1) can also be produced using the summit cannon. For example, if we further increase the barrel speed to around, say, 9 km/sec (5.59 miles/sec), this has the effect of raising the height of the ball's trajectory on the other side of the globe, producing an elliptical trajectory (Figure 2.1). Since this is a closed trajectory, the cannonball will come back to haunt the cannon crew, about $2^3/_4$ hours after the projectile is fired, in this case. Note that the ball always returns to the low point on the orbit, the summit cannon, which is referred to as the orbit *perigee*. The high point, on the other side of Earth from the mountain, is called the orbit *apogee*. These are rather strange terms, but as the topic of orbit dynamics has been with us for so many years, a lot of wonderful terminology has come to us from history, as we will see later. Getting back to our cannon, further ramping up the barrel speed will result in higher and higher apogees, giving more and more elongated ellipses. Eventually, the apogee height will effectively reach infinity, an extremely long way away, and then the trajectory becomes an open parabola (Figure 2.1).

If you ask the cannon crew to check the barrel speed, the crew members will tell you that the parabolic trajectory occurred at around 11 km/sec (6.84 miles/sec). If you also recall the discussion about the parabola in Chapter 1, it is the trajectory that results in escape from Earth's gravity with the minimum energy given to the cannonball. The ball flashes out of the cannon at huge speed, but this energy is consumed by the gravity field as it climbs away from Earth, and when it reaches infinity it effectively has no energy left to go anywhere, that is, it has zero speed. Any further increase in the barrel speed of the cannon will result in the ball's trajectory being a hyperbola (Figure 2.1). If you recall, this gives the ball sufficient energy to escape Earth's gravity, with some left over to give it a constant speed once it has reached a great distance from Earth.

While Newton's cannon is helpful in revealing the nature of orbital motion, as you have probably guessed it does not have much to do with the realities of launching current spacecraft into orbit. This is done using launch vehicles, and we shall discuss in Chapter 5 how these are related to Newton's cannon.

Weightlessness

Now that we have a good feel for the nature of orbital motion—essentially a spacecraft is in a state of free-fall under gravity—we can also achieve a similarly good understanding of the phenomenon of weightlessness.

Weightlessness is something we see routinely on news coverage of manned space missions. (In this book I use the phrase *manned space missions* to mean flights involving people—both men and women. I know that the phrase may not be quite politically correct, but I dislike the other possibilities, such as "crewed" missions or "peopled" missions.) We have become familiar with crew members floating about their space ships, performing tricks such as swallowing floating globules of water, which would of course be impossible back on Earth. Despite this familiarity, however, there are again misconceptions about the nature of weightlessness, but it can be easily understood in terms of objects—spaceships, astronauts, and globules of water—free-falling together in a gravity field.

The key to understanding is an appreciation that all objects, independent of size and mass, fall with the same acceleration in a gravity field. The first statement of this principle is attributed to Galileo Galilei, who was born near the city of Pisa in 1564. To prove it, he is said to have dropped a cannonball and a wooden ball of the same size from the top of the famous leaning tower to demonstrate that the two balls would impact the ground at the same time, despite their different weights. Unfortunately, it is agreed by historians that this rather splendid story is of doubtful authenticity.

A much better demonstration was performed on the moon's surface in July 1971, by Apollo 15 astronaut Dave Scott, who dropped a feather and a hammer together to see which of them would reach the lunar dust first. Since all three astronauts on this mission were serving members of the United States Air Force, the landing module was named Falcon, after the mascot of the U.S. Air Force Academy. The feather had to be that of a falcon, a detail that is of course entirely immaterial! You can perform this experiment now—if you happen to have a feather in one pocket and hammer in the other—but I think you can guess the outcome. Clearly the feather, being much lighter than the hammer, will hit the ground some time after the hammer in apparent contradiction of Galileo's assertion that all things fall with the same acceleration. However, the experimental method in this case is flawed; there is the unfortunate presence of air in the room—fortunate for you, but not for the experiment! In the lunar surface experiment there is no air to influence the motion of the feather, and the feather and the hammer hit the dust at the same moment, giving a convincing demonstration that objects do fall at the same rate in a gravity field.

We can gain an understanding of weightlessness in orbit in terms of the spacecraft, the astronaut, and all other free objects inside the vehicle all falling together with the same acceleration in Earth's gravity field.

To consolidate this idea, we can attempt to do an experiment on the ground to reproduce the effects of weightlessness by replacing our spaceship with an elevator in a very tall building. Strangely, the elevator is equipped with a weighing machine, as shown in Figure 2.2a. When we enter the elevator, while it is stationary, we can climb on the weighing machine, and we know that it will register our normal weight. We also have sufficient experience of riding elevators to know that, if we were to press a button to go up, we will feel heavier while the lift cable is accelerating the elevator upward—the weighing machine will register this increase in weight (Fig. 2.2b). However, the part of the experiment to simulate weightlessness (Fig. 2.2c) is not to be recommended, as it involves cutting the elevator cable while disabling the elevator breaking system! In this case, the elevator and all objects within it will free-fall under gravity with the same acceleration, giving the same effects of weightlessness as seen on a spacecraft but for a rather shorter period of time!

Interestingly, a number of research laboratories around the world offer such a facility commercially (called a drop tower), in which hardware experiments—but not people!—are dropped to produce brief periods of weightlessness. Note that, in this discussion, it is important to make a clear

Figure 2.2: (a) Elevator stationary. (b) Acceleration of elevator upward causes increase in weight. (c) Elevator in free-fall produces weightlessness.

distinction between weight and mass. Our unfortunate elevator rider in Figure 2.2 may have a mass of, say, 80 kg (175 lb), which remains constant throughout his scary adventure, but as we have noted his weight varies considerably depending on the state of motion of the elevator. Mass, according to Isaac Newton, is a fixed attribute of an object that characterizes its inertia; massive objects like pianos require a significant push to get them moving, whereas smaller objects require much less effort.

This difference between mass and weight also had a surprising consequence for the Apollo moon-walking astronauts, who found they fell over rather more often than they were anticipating. A typical astronaut's mass, including space suit and backpack, was on the order of 130 kg (285 lb), but of course their weight in the lunar surface gravity was around one sixth of their Earth weight. This difference meant that the friction between their boots and the moon's surface was similarly reduced to one sixth of that on Earth. As they moved around on the moon's surface, sometimes quite rapidly, they had less contact friction with which to manage their significant mass—with some interesting results!

Spacecraft Mission Analysis

After the historical perspective of Chapter 1, and the earlier sections of this chapter, we now move on to begin to tell the story of modern spacecraft design. In the remainder of this chapter and the subsequent two chapters, we continue the theme of orbits, but in the context of spacecraft in orbit around Earth, or around another planet, or indeed around the Sun. *Spacecraft mission analysis* is a rather fancy term that spacecraft engineers use to describe the design of the orbital aspects of a spacecraft mission. On any spacecraft project there will be a team of people tasked with this job, which involves things like selecting the rocket that will launch the spacecraft, selecting the best orbit for the spacecraft to achieve the objectives of its mission, and determining how the spacecraft will be transferred from launch pad to final orbital destination.

Orbit Classification

To discuss the types of closed Earth orbits that are commonly used by spacecraft operators, we need to consider the characteristics of typical orbits that uniquely distinguish one orbit from another. Principally, these distinguishing features are shape, size, and inclination.

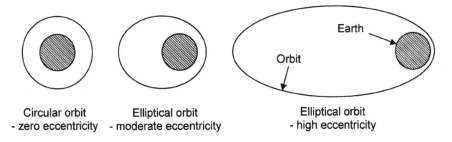

Figure 2.3: Shape is a principal distinguishing characteristic of orbits. The degree of elongation of an orbit is defined by its eccentricity.

For closed orbits, the relevant *shapes* are circles and ellipses. Of course, some ellipses are more elongated than others, as shown in Figure 2.3, and this degree of elongation is referred to as *eccentricity*, with high eccentricity orbits being more elongated.

Similarly, *size* is an easy idea, being defined by the orbit height. More precisely, a circular orbit will be defined by its radius, measured from Earth's center, or by its altitude above Earth's surface, as shown in Figure 2.4a. For elliptical orbits, the overall size of the orbit can be defined in terms of the distance between perigee and apogee (Fig. 2.4b). The perigee and apogee points may also be pinned down by their respective distances from Earth's center or surface.

The third principal characteristic, *orbital inclination*, essentially defines the orientation of the orbit plane with respect to Earth's equator, as illustrated in Figure 2.5. The orbital inclination is defined as the angle between the orbit plane and the equatorial plane, measured at the ascending node of the orbit. Again in terms of the jargon, a *node* is simply a

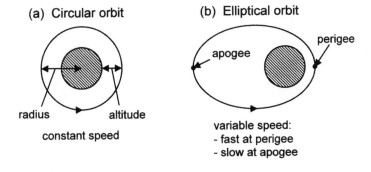

Figure 2.4: The size of an orbit is a principal characteristic, and this is defined by the orbit's altitude above Earth's surface, or its distance from Earth's center.

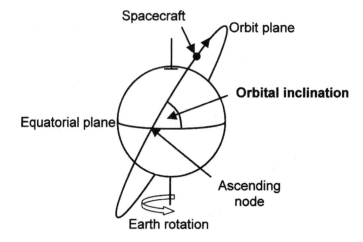

Figure 2.5: The angle between the orbit plane and the equatorial plane is called the orbital inclination. This is the third principal distinguishing characteristic of an orbit.

point on the orbit where the spacecraft crosses the equator, and an *ascending node* is one where the spacecraft is traveling from south to north. Looking at Figure 2.6a, we can see that an orbital inclination of 0 degrees gives an equatorial orbit, that is, one that overflies the equatorial region only. Conversely, an orbital inclination of 90 degrees gives an orbit plane perpendicular to the equatorial plane, as shown in Figure 2.6b. This type of orbit is referred to as a polar orbit. Of course, the orbital inclination may take any value between 0 and 180 degrees; a value of about 45 degrees is illustrated in Figure 2.6c.

Another property of the orbit that is of interest, implied in Figure 2.4, is the *orbital speed* with which spacecraft move along their orbital path. This is not a principal characteristic, but an attribute that arises as a result of the

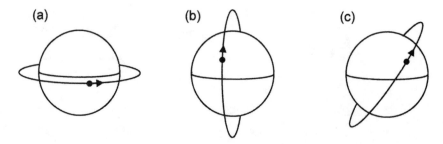

Figure 2.6: Orbits of differing orbital inclinations. (a) An equatorial orbit. (b) A polar orbit. (c) An orbit with an inclination of about 45 degrees.

orbit shape and size. The mathematics tells us that in a circular orbit, the spacecraft's speed is dependent on the mass of the central body and the orbit height. Given that we are considering Earth orbits, the mass of the central body, Earth, is of course constant, so the spacecraft's speed then becomes dependent only on the orbit height. A spacecraft in a circular orbit at particular altitude will move at a precisely defined speed. For example, in a 200-km (124-mile) altitude circular orbit, the spacecraft moves at 7.78 km/sec (4.84 miles/sec), as we have already seen in our discussion earlier of Newton's cannon. This is a low Earth orbit (LEO). The rule is that as the circular orbit height increases, the orbit speed decreases; for example, a spacecraft in a 10,000-km (6200-mile) altitude circular orbit will travel at around 5 km/sec (3 miles/sec).

In an elliptical orbit, the spacecraft speed along its trajectory is slightly more difficult to quantify, as the mathematics are a little more involved. But it can be understood easily in terms of the sharing of the spacecraft's energy between height and speed. As the spacecraft's altitude increases, its energy is sapped by its climb out of the gravity field and it slows down. At the apogee point of an elliptic orbit, the spacecraft speed will be lower than its speed at perigee. We have seen this already, expressed in a rather geometrical (17th century) way by Kepler's second law of planetary motion (see Chapter 1). A good parallel to help remember the variation in speed in elliptical orbits is bike riding on hilly terrain; your speed in the valleys is much higher than when climbing to the high points, for the same reasons of converting height into speed, and vice versa, as you ride.

Popular Operational Orbits

Now that we have a grasp of the three principal distinguishing characteristics of orbits—shape, size and orbital inclination—we can begin to look at the Earth orbit types that are most commonly used by spacecraft operators. Obviously, if we allow all possible variations in these three characteristics, then there is an infinite number of resulting Earth orbits to choose from! The popular orbits that we are about to introduce, therefore, are a small subset of this vast number of possibilities, and these are widely used simply because they have useful properties that enhance the performance of scientific and applications spacecraft. In writing this chapter, I found it difficult to decide what to include and what to leave out. No doubt other experts would say, "Well, what about such and such an orbit, which is often used for this or that?" I guess the reader has to accept that sometimes things are simplified and generalized a little to aid clarity.

Bearing in mind these comments, five types of popular operational orbit are identified, and these are summarized in the following box for quick reference.

Some popular operational Earth orbits

1. **Low inclination LEO**
   A circular Low Earth Orbit with an orbit plane varying from equatorial up to about 50° inclination.
2. **Near-polar LEO**
   A circular Low Earth Orbit with an orbit plane inclined near 90°.
3. **HEO**
   A Highly Eccentric Orbit.
4. **GEO**
   A circular, equatorial orbit at a height where the orbit period is 1 Earth day. This is referred to as a Geostationary Earth Orbit.
5. **Satellite constellation orbits.**
   A network of usually identical circular, inclined orbits, often accommodating a large number of satellites.

Low-Inclination LEO

This is a circular low Earth orbit, with an orbit plane that is near-equatorial (Fig. 2.7). However, the simplicity of this statement is deceptive, and what we mean by "low" and "near-equatorial" requires qualification.

Figure 2.7: Low-inclination low Earth orbit (LEO). Large vehicles, such as space shuttles and space stations, are often accommodated in this type of orbit. (Image courtesy of the National Aeronautics and Space Administration [NASA].)

Surprisingly there is much debate among experts about the meaning of the word *low*, but my working definition is altitudes below about 2000 km (1240 miles). The phrase *near-equatorial* similarly gets a wide interpretation, meaning orbit planes ranging up to around 50 degrees in orbital inclination. The kinds of spacecraft found in this type of orbit are often large, that is, massive, manned vehicles such as shuttles and space stations, or large unmanned spacecraft. An example of this class of unmanned vehicle is the well known Hubble Space Telescope, which is about the same size and mass as a double-decker bus—around 11,000 kg (24,000 lb). The mass of space vehicles and the type of orbit in which they are accommodated are related. As we will explain in more detail in Chapter 5, it is much easier to launch large spacecraft into low, near-equatorial orbits.

Also, given that the plane in which the planets orbit the Sun is close to Earth's equatorial plane, spacecraft destined to probe distant planets are often launched into a near-equatorial LEO. This is then used as a kind of *parking orbit*, to check out the spacecraft's onboard systems, before a rocket is fired to take the probe to its ultimate destination.

Near-Polar LEO

This orbit is used mostly by operators of Earth observation and surveillance spacecraft (Fig. 2.8). It is a popular operational orbit, particularly at altitudes in the region of 700 to 1000 km (435 to 620 miles), and this is mainly driven by a need to get a global perspective on environmental issues such as climate change. Consequently, many national and international space agencies have launched (and are planning to launch) an armada of spacecraft equipped with powerful instrumentation directed downward to the Earth's surface. Earth observation also has a military dimension, and many military agencies are launching surveillance satellites to gain the new military "high ground." It is perhaps not well known that the biggest spender on space in the world is the U.S. Air Force, and details of most of their spacecraft and activities are classified. However, to get a feel for the capabilities of their optical surveillance satellites, you have to imagine a spacecraft with similar imaging power to the Hubble Space Telescope, but directed down instead of up!

It is easy to see why near-polar LEOs are good for Earth observation. Figure 2.8 demonstrates that there is potential for our spacecraft to see most of the planet's surface if we wait long enough; this is called *global coverage.* As our spacecraft orbits once every 100 minutes (typically), and Earth rotates once every 24 hours beneath the orbit plane, the spacecraft operators can image most targets of interest worldwide within a day or two. The targets of interest can vary substantially in character, from the health of a crop of maize to tank movements on a battlefield.

Figure 2.8: Near-polar LEO. This is commonly used by Earth-observation satellites, like the Landsat spacecraft shown. (Image courtesy of NASA.)

If we compare the near-polar LEO with the low inclination LEO, we can get a good idea of why the near-polar orbit is so well suited to Earth observation missions. It is obvious from Figure 2.7 that if we launched an Earth observation satellite into a low-inclination orbit, we would get a good look at the near-equatorial regions of Earth, but not much else.

HEO

The geometry of a typical highly eccentric orbit is shown in Figure 2.9, which is inherently useful for a variety of missions.

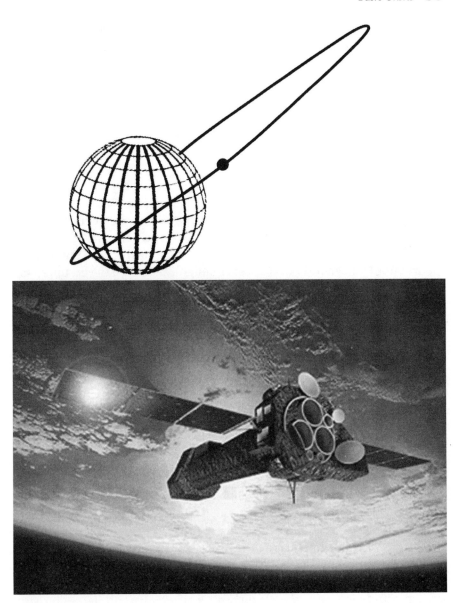

Figure 2.9: The highly eccentric orbit (HEO). One type of mission that it accommodates is astronomical observatories, such as the XMM Newton X-Ray Telescope. (Image courtesy of the European Space Agency [ESA].)

The HEO has accommodated many scientific spacecraft, for example, the European Space Agency Cluster mission, dedicated to exploring Earth's magnetic field, and the energetic atomic particles that are trapped within it.

The source of these particles is the solar wind, a stream of high-speed ions—atoms stripped of their electrons—emanating from the Sun. When these encounter the magnetic field of Earth, some of them are trapped in near-Earth space to form doughnut-shaped regions filled with energetic particles (see Chapter 6). These regions were named the Van Allen radiation belts in 1958 after their discoverer, and the particle radiation they contain is hazardous to both man and spacecraft systems. Understanding this hazard has been an important endeavor, and the HEO provides the best means of doing this, as a spacecraft in this orbit is able to sample the magnetic field and particles over a wide range of altitudes on each orbit.

The HEO is also a popular orbit for space observatories. An observatory at the apogee of a HEO has good *sky viewing efficiency*, as the distant Earth obscures only a small part of the sky. Also the relatively slow speed at apogee means the spacecraft spends most of its time there, providing good opportunities for extended periods of communication with the ground. This allows ground operators to command the telescope and receive its data as if it were effectively on the ground in a dome next to the control room. This way of operating is referred to as *observatory mode operation* and is an important attribute for space telescopes. Also the high apogee of the HEO means that the observatory spends most of its time above the Van Allen radiation belts, which is beneficial for some instruments that cannot operate in a high radiation environment.

The HEO has also been used extensively as a communications orbit, mainly by the former Soviet Union and by Russia today. A HEO inclined at 63 degrees to the equator, with an orbit period of 12 hours, is called a *Molniya* (Russian for "lightning") orbit after a series of communications satellites accommodated in this orbit. The Soviet Union began to use this orbit in the 1960s for communications between ground sites at high northern latitudes, by positioning the apogee of the orbit above the Northern Hemisphere. The low speed of the spacecraft in the apogee region means that it spends the majority of its orbit period high in the sky above these northern regions, giving an opportunity for extended, uninterrupted periods of communication with terrestrial users. Between 1964 and 1998, around 170 spacecraft were launched into *Molniya* orbits to provide telephone communication and satellite TV to high-latitude regions bordering the Arctic Ocean.

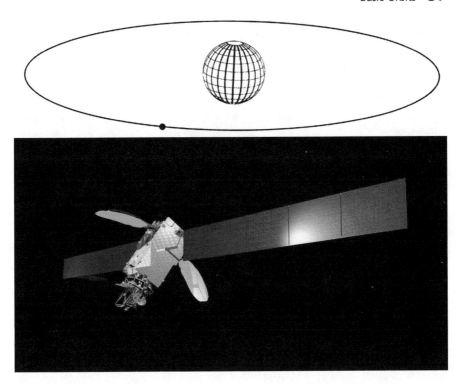

Figure 2.10: The geostationary Earth orbit, shown to scale. There are many communication satellites in this orbit, such as the Intelsat spacecraft illustrated. (Image courtesy of EADS Astrium.)

GEO

The geostationary Earth orbit is a widely used operational orbit, mostly for communications, but also for scientific and Earth observation satellites (Fig. 2.10). An example of a GEO orbit Earth observation satellite is Meteosat, which provides those impressive weather pictures we see each evening on the television weather forecast. The invention, if such it can be called, of the GEO is attributed to the science-fiction author Arthur C. Clarke in 1945. Unfortunately, he failed to patent the idea; if he had done so, he would probably be very rich today! The reason for the popularity of the GEO as an operational orbit is its unique characteristic that satellites in this orbit appear to be stationary when seen from Earth's surface—hence the name. To achieve this, the orbit needs to be circular and equatorial, but in addition the orbit height has to be such that the spacecraft orbits Earth in the same time as it takes for Earth to rotate once on its axis.

There is often confusion about the meaning of the term *geosynchronous orbit* (GSO) and how it relates to the GEO. GSO is the name used for any orbit that has an orbital period equal to one Earth rotation, so the GEO is a special case of the GSO. There are obviously a whole bunch of GSOs with a 1-day period, but having an elliptical shape, or a plane inclined to the equator, or both. The important distinction is that a spacecraft in one of these GSOs does not appear stationary relative to a ground-based observer.

Usually the orbit period of a GEO is said to be 24 hours, but it is actually a little shorter than that—23 hours and 56 minutes. The day on which we base our calendar is the familiar 24 hours, which is called the *solar day,* and this is the time it takes for Earth to rotate once with respect to the Sun. If the Sun is precisely due south (or due north if we live in the Southern Hemisphere) and we measure the time it takes for it to return to the same position in the sky the next day, we will find it to be the familiar 24 hours. The period of the GEO satellite, 23 hours 56 minutes, is called the *sidereal day,* which is the time it takes for Earth to rotate once with respect to the distant stars. The reason for the difference is Earth's orbital motion around the Sun; because of this, the Sun appears to move relative to the stars. As Earth rotates, it takes 23 hours and 56 minutes to do one revolution with respect to the stars, and then it has to rotate for an extra 4 minutes to catch up with the Sun, as the Sun's position has changed from the day before.

Getting back to our GEO spacecraft, we can calculate the orbit height corresponding to this orbit period using Kepler's third law of planetary motion (see Chapter 1). If we do this, we get a precise altitude for our GEO of 35,786 km (22,237 miles). If we have a circular, equatorial orbit at this height, a satellite initially positioned above a particular geographical feature on the equator will remain above that feature as it orbits; it appears to stand still in the sky from the point of view of someone on the ground.

This property is the key to its popularity. It makes communication with the spacecraft easier, as you don't need to track the satellite with your dish antenna—you just point it in a fixed direction. And of course it also means that the communications link with the spacecraft is uninterrupted. This is a familiar idea, as evidenced by the large number of small satellite TV receiving dishes we see bolted to the exterior walls of houses, staring fixedly at a particular point in the sky where the service provider's invisible GEO satellite resides.

The GEO orbit is most commonly used by communication spacecraft, and there are literally hundreds of active communication satellites (comsats) on the GEO arc. People routinely use this technology day to day, without really noticing, which is of course the way it should be. If I pick up the phone to say, "Hi, this is Graham" to a friend on another continent, then my electronic voice will transit over land lines or microwave links to the nearest satellite ground station, where it will be transmitted into the sky to a GEO comsat by a large fixed dish antenna. This signal will be received and amplified by the spacecraft, and then transmitted down to another ground station in the region of my friend's home, ultimately arriving at his telephone handset. When he responds by saying, "Oh hello! How are you?," the whole process begins again in the reverse direction—amazing technology that is transparent to the user!

The nature of the GEO arc is that it is literally a one-dimensional line in space, and as such it is a limited natural resource that needs to be protected and managed for the future, like any other. Unfortunately, as well as all the active comsats on GEO, there are many defunct spacecraft that are essentially debris polluting the orbit. Because of the pressure of use on the GEO arc, spacecraft operators are now expected to boost their comsats to a higher graveyard orbit—200 or 300 km above GEO—when they reach the end of their operational life.

Satellite Constellation Orbits

The orbits associated with a satellite constellation are usually a network of identical inclined circular orbits, often accommodating a large number of satellites. A typical constellation geometry is illustrated in Figure 2.11, where the black dots represent the orbiting satellites comprising the constellation. Constellations have been most commonly used over several decades for satellite navigation (satnav), which uses satellites to determine your position on the ground (or on the ocean, in the air, or wherever you happen to be). More recently, constellations have been used for satellite communications, and there is currently an interest in using them for Earth observation as well.

Perhaps the best known example of a constellation is the global positioning system (GPS) navigation system. Navstar GPS satellites (see Chapter 1) are operated by the U.S. Department of Defense, mainly for use by the U.S. military. However, satnav in automobiles is becoming commonplace, as well as in leisure activities such as hiking and sailing, giving the user's position with an accuracy on the order of 10 m (32 feet). To triangulate a user's position on the ground, the receiver needs to access signals from at least four GPS satellites simultaneously. To make this work, the constellation must be designed so that the user can see at least four GPS satellites from any location on Earth's surface at any time. This ground coverage requirement leads to the design of the geometry of the satellite constellation. In this case, the required ground coverage is achieved by the operation of 24 satellites in the constellation. The resulting geometry of the constellation consists of six circular orbit planes at 20,200 km (12,500 miles) altitude, spread out around the equator. Each orbit plane is inclined at 55 degrees to the equator and accommodates four satellites. Figure 2.12a shows

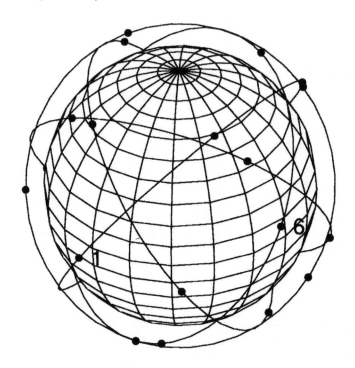

Figure 2.11: An illustration of a typical LEO constellation geometry for a communications system. The black circles represent the orbiting satellites that comprise the system.

the GPS constellation geometry, illustrating well the old saying that a picture is worth a thousand words! A typical GPS satellite configuration is shown in Figure 2.12b.

However, the Navstar GPS system is unlikely to be the future of space-based navigation systems because of its military ownership. Quite reasonably, the Department of Defense reserves the right to limit the signal strength, to erode the positional accuracy of the GPS system, and to shut down public access to GPS completely in times of military conflict. This political aspect of the GPS system has overridden the amazing technological benefits, and means that civilian agencies have been reluctant to embrace space-based navigation systems wholeheartedly. Just think how useful satnav would be if fully utilized for things like air traffic control, particularly now when the density of air traffic is growing at such a phenomenal rate. To overcome these political difficulties, the European Union has proposed launching a new satellite navigation constellation called Galileo, which will have civilian ownership. This system, comprising 30 satellites, is due to be operational in around 2012, and should allow space-

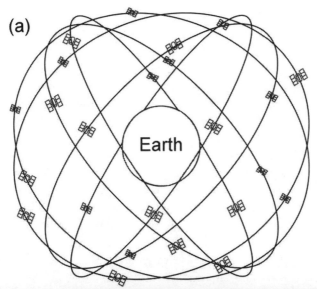

Figure 2.12: (a) Schematic of the global positioning system (GPS) Navstar constellation geometry. (b) A typical GPS satellite configuration. (Image courtesy of Lockheed Martin.)

based navigation techniques to become more fully integrated into all aspects of human activity where high precision positioning is required.

Satellite constellations have also been proposed for communications, and in this application low-altitude orbits are required. An example of this type of LEO constellation is shown in Figure 2.11. We have already discussed the usefulness of the GEO orbit for global communications, so why is there a need to propose a whole new way of doing the same thing? Well, the spur for this development is the craving we have for communications using small hand-held terminals, or, put another way, mobile phones. This love affair with mobile phone technology has created a huge expansion in terrestrial cell phone networks. As we roam around the planet, the communications with our friends or business associates is achieved by the links between our phones and a network of fixed, ground-based, mobile phone antenna masts. This works well most of the time, as the cell phone network operators have located the fixed phone masts in places where population is concentrated. This is, of course, driven by market forces. However, we all know that if we go for a hike in a remote mountain region and we wish to contact someone, the mobile phone will probably not work as we are out of range of the terrestrial network of mobile phone masts. What do we do if we find ourselves on top of Mount Everest and wish to contact a friend who is traversing the Gobi Desert? Yes, I know—a rather unlikely scenario—but it makes the point. In this situation we can adopt a satellite solution to the problem of mobile communications. Instead of having a network of fixed masts on the ground, we can establish a network of satellites in the sky that perform the same function. If we design the orbital geometry of the sky network or satellite constellation appropriately so that we always have line of sight to at least one satellite member of the constellation, then truly global mobile phone communications become possible. The sort of constellation geometry we see in Figure 2.11 is again driven by the coverage requirement that we need to see at least one satellite from all terrestrial locations, and at all times.

Why do we need a LEO constellation of satellites for mobile phone communications? Why can't we just talk through the existing network of GEO satellites? The simple answer is that the GEO satellites are just too far away. If our mobile phones had enough microwave transmission power to reach the 36,000 km (22,400 miles) or so to GEO, then they would literally fry our brains in the process—quite a major physiological constraint on the technology!

If we take account of all these various considerations, then mobile phone constellations usually end up comprising a network of near-polar LEOs. Perhaps the best-known example of this type of communications constellation is the Iridium system. The original proposal for this constellation was to have 77 active satellite members, equal to the number of electrons of an iridium atom, thus giving rise to its name. Subsequently this was reduced to 66 active satellites in a network of near-polar, circular LEOs at a height of approximately 780 km (485 miles). Because of competition from terrestrial mobile phone networks, Iridium has had a checkered commercial history, which has inhibited the growth of space-based mobile communications. However, if the economics can be gotten right, then, in the words of the ad, the future's bright for this type of space application!

The third main application of constellations is Earth observation, which is currently the least well-developed, although there may be military developments of which I am ignorant! When we briefly discussed Earth-observation satellites in the near-polar LEO section, we commented that the spatial resolution of current imaging payloads were amazing. Objects the size of a fraction of a meter can be easily seen from orbit with the right payload equipment, provided that the ground is not obscured by cloud. However, one problem with conventional Earth-observation spacecraft is their temporal coverage. As the satellite orbits every 100 minutes or so, and Earth rotates once every 24 hours beneath the orbit plane, it may take a while for the spacecraft to have an opportunity to over-fly and image a particular ground target of interest. The temporal coverage is not good using a single satellite. Overcoming this limitation is particularly important in military operations, where uninterrupted strategic battlefield information may be a requirement. Various civilian applications, such as disaster monitoring, would also benefit from improved temporal coverage, and using constellations of Earth-observation satellites is a way of achieving this. In principle, a continuous line of sight to a ground target of interest is achievable using a constellation provided that sufficient imaging satellites are launched.

From the above discussion, it is obvious that there a lot of advantages to using satellite constellations. Another one that we have not mentioned is *graceful degradation*. If you launch one satellite to provide a service, such as communications or Earth-observation, and it suffers a serious system or payload failure, then the service it provides is abruptly interrupted. However, if the function of providing the service is distributed among a large number of constellation satellites, then clearly the failure of one satellite means that the service may be compromised a little, but nevertheless it can be maintained. This characteristic of a more robust operation, associated with constellations, is particularly important in military space activities, where an adversary may be actively seeking to interrupt normal service!

Finally, constellations have the disadvantage of the cost of manufacturing, launching, and operating the many satellites in a constellation system, which is much higher than the cost of a single satellite system. However, as we have seen with navigation and mobile communications services, this burden of the cost has been taken on by the operators, as the many-satellite attribute of a constellation is essential in achieving the objective.

Choosing the Best Orbit

The five types of orbit we have discussed are popular with spacecraft operators, but how do we select the best of the huge number of possible orbits to choose from for a particular spacecraft mission? This question is central to the activities of the team of project engineers tasked with the job of the spacecraft mission analysis.

The quick answer is that the spacecraft needs to be in the right place, that is, the right orbit, so that the spacecraft payload can most effectively achieve its mission objectives. We need to pause a moment to reflect on this concise but not so simple statement. First, what is the *spacecraft payload*? Essentially it is the part of the spacecraft that fulfills the mission objectives—the business end of the spacecraft. For example, the payload of an Earth-observation satellite will be the camera instruments used to acquire the image data, or the payload of a communications satellite will be all the telecommunications equipment and antennas needed to maintain the desired communications service. Second, what is a *mission objective*? This is the purpose behind the whole project, its *raison d'être*. Some typical spacecraft mission objectives are the following:

1. The provision of high-resolution imagery of Earth with global coverage
2. The provision of a telecommunications service for the Australasian region using large, fixed ground antennas
3. The acquisition of high-resolution astronomical imagery

The process of linking the mission orbit selection with the mission objective usually involves the following steps:

- Definition of the spacecraft mission objective: the formulation of a precise statement defining the prime purpose of the spacecraft.
- Choice of payload instruments or equipment, usually done by a group of experts who can produce a detailed specification of the payload hardware required to achieve the objective.
- Development of payload operational requirements: How does the payload hardware need to operate to best achieve the objective? This

includes where the payload needs to be physically located to maximize its effectiveness.

- Finally, the consideration of the payload's location leads naturally to the selection of an appropriate, or even optimal mission orbit.

All this may sound rather formal and complicated, but we can return to our example mission objectives above to show that sometimes the process can be rather straightforward.

If we look again at mission objective 1, above, related to Earth observation, the requirement for *high-resolution* imagery of Earth means that the imaging payload instruments need to be close to their terrestrial *targets of interest*, which in terms of an orbit translates to a LEO. The need for *global coverage* means that our LEO must be near-polar in inclination to provide the instruments the opportunity to "see," after some period of time, the Earth's entire surface. Without too much difficulty, the use of a near-polar LEO (see Fig. 2.8) seems the obvious choice in this case. Similarly, it is easy to see that the choice of "best" mission orbit for mission objective 2, above, is a GEO.

These two examples illustrate well the process of how the mission objective drives the choice of mission orbit for a particular spacecraft project. However, just to muddy the waters a little, we can look at an example that shows that sometimes the choice of the mission orbit is not quite so obvious. The third example, mission objective 3, above, relates to the operation of an orbiting astronomical observatory, and if we look at the orbits of such spacecraft in current operation we find them in a variety of orbits. For example, the Hubble Space Telescope can be found in a LEO, the XMM Newton X-Ray telescope orbits in a HEO, and the Hipparcos observatory was designed to fly in GEO (although, due to a rocket engine failure, it did not make GEO and went into a HEO instead!). This variety suggests that the choice in this case is perhaps not quite so simple.

In cases such as this, a more detailed analysis is required to select the orbit, involving consideration of both spacecraft payload and system requirements which influence the decision. This kind of process is illustrated in Table 2.1, which is a simplified version of a trade-off table that engineers might use to help make an orbit selection for a space observatory. In a typical trade-off process, the objective is to make a choice among a number of different options; in this case the orbit options are LEO, HEO, and GEO (the right-hand columns of the table). To make this choice, a number of criteria or trade-off parameters, are specified (in the left-hand column of the table) against which the options are judged.

In our choice of parameters, three are related to the payload (telescope) operation, and three are related to spacecraft system operation. You may

Table 2.1: A simple orbit trade-off table for an orbiting astronomical observatory

Parameter	Type of parameter		Favoured orbit		
	Payload	System	LEO	HEO	GEO
Observatory mode operation (duration of ground communications link)	✓			✓	✓
Uninterrupted source observation	✓			✓	✓
Sky viewing efficiency	✓			✓	✓
Radiation exposure		✓	✓		
Ease of orbit acquisition		✓	✓		
In-orbit repair and maintenance		✓	✓		

recall, from the discussion about observatories in HEOs earlier, the explanation of *observatory mode operation* and *sky viewing efficiency*. A third payload parameter, *uninterrupted source observation*, is introduced, which relates to how long the telescope can point at a particular nebula or galaxy without interruption. To maximize sensitivity, telescopes often operate in a kind of time-exposure mode, where they point at the object of interest for long periods of time to collect as much light as possible. When you are looking at extremely distant objects at the edge of the universe, collecting every photon of light counts. If you think about an observatory in a LEO with a period of around 100 minutes, this kind of operation is difficult to achieve because Earth can get in the way of the telescope for about 30 minutes on each orbit revolution. Table 2.1 shows that high orbits—HEO apogee and GEO—are favored when we consider the payload parameters.

On the other hand, when we look at our system-related parameters, the LEO is the preferred option. The *radiation exposure* in LEO is less severe than in the higher orbits, so the reduced degradation of the spacecraft systems caused by particle radiation favors the LEO. The *ease of orbit acquisition* parameter relates to the amount of propulsive effort required to reach the mission orbit. Again the LEO orbit is favored as it requires the least amount of rocket fuel to get there, compared to the higher orbits. And savings in fuel mass in getting to your mission orbit can be usefully invested in increasing the payload mass, resulting in an improvement in the overall effectiveness of the spacecraft in achieving its objective. The third parameter, *in-orbit repair and maintenance*, is a means of effectively increasing the mission lifetime of the observatory. This maintenance is usually done by space-walking astronauts, and since manned vehicles can only visit LEOs routinely, this form of maintenance can only be performed on LEO spacecraft.

After all that, when we look at the checked boxes on the right-hand side of Table 2.1 we may be disappointed to find that the choice of mission orbit for our observatory is still unresolved! In truth, the process illustrated above is too simple to make any real progress, but it is useful in illustrating the trade-off process. In a real project situation, the trade-off exercise would be much more finely tuned to the specific spacecraft characteristics. Also, some of the parameters that are considered to be particularly important would be weighted in the trade-off process. For example, in choosing the best orbit for the Hubble Space Telescope, the issues of ease of orbit acquisition and in-orbit maintenance were paramount. The sheer size and mass of the telescope meant that LEO was the only real option for the orbit. Also, the desire to extend its useful lifetime by the process of in-orbit maintenance requires the use of astronaut repairmen taken aloft by the U.S. Space Shuttle, so the telescope's orbit is again constrained as the shuttle is unable to operate above LEO. As a consequence, the Hubble Space Telescope ended up in a LEO, even though we can see from Table 2.1 that it is not the best orbit for a space observatory!

In this chapter we have come a long way from Newton's cannon to popular operational orbits for modern scientific and applications spacecraft, and in the process have acquired an understanding of the nature of orbital motion. In Chapter 3, which discusses real orbits, we delve into this topic a little more deeply, to discuss the mysteries of orbital perturbations and how they influence the process of mission analysis. Give it a try, but if you find it hard going, skip it and move on. The rest of the book is not crucially dependent on the content of Chapter 3.

Real Orbits

Ideal and Real Orbits

THE orbits that we have discussed in the preceding two chapters are often called Keplerian orbits, named after Johannes Kepler, and sometimes referred to as *ideal orbits*. If we imagine a satellite in orbit around the Earth, and the only force acting upon it is that of a perfect inverse square law gravity field (see Chapter 1), then the resulting orbital motion is described as ideal. The main distinguishing feature of an ideal orbit is that the attributes defining the orbit—its shape, size, and orbital inclination (see Chapter 2)—all remain fixed; that is to say, they do not change with time.

In contrast to an ideal orbit, the defining attributes of *real orbits* do change, and this is due to the effects of *orbital perturbations*. This is just a fancy phrase to describe forces that act on the satellite, in addition to the inverse square law of gravity, that cause its path to differ from an ideal circular or elliptical orbit. When the mission analysis team designs the orbital aspects of an Earth orbiting mission, it has to take these aspects into account. There are a number of different sources of perturbations, and this chapter discusses the most common ones.

The gravity field of the Earth is very close to being described by Isaac Newton's inverse square law, although there are small departures from this. The dominant part of Earth's gravity field, described by the inverse square law, is sometimes referred to as the *central gravity field*. The small departures are sometimes called *gravity anomalies* and are a form of perturbation that causes the satellite to deviate from ideal orbital motion. Speaking more generally, the effects of orbit perturbations from all sources are small compared to the central gravity field. As a consequence, real orbits are basically the same shape as ideal orbits—circles and ellipses—but their shape, size, and orbital inclination change slowly over time.

G. Swinerd, *How Spacecraft Fly: Spaceflight Without Formulae,*
DOI: 10.1007/978-0-387-76572-3_3, © Praxis Publishing, Ltd. 2008

Earth Orbit Perturbations

For Earth-orbiting spacecraft, there are four main sources of orbit perturbation that can have a significant effect on the orbital motion:

1. Gravity anomalies
2. Third-body forces
3. Aerodynamic forces
4. Solar radiation pressure

In what follows we summarize how these work, and discuss in what way they affect the characteristics of the orbit. There are many other perturbation forces of lesser magnitude that influence the motion; in fact, the list of possible perturbations is long. For a particular spacecraft, the choice of which perturbations to include in controlling and operating the spacecraft depends on how precisely we need to know where the spacecraft is located in space. For example, the position of some current Earth-observation satellites needs to be known to an accuracy of a few centimeters in order to get the maximum amount of useful science from the payload. In this case, a long list of orbit perturbations must be taken into account to determine the position of the spacecraft so accurately. On the other hand, to operate a geostationary orbit communications satellite successfully, it may be necessary to know only that it is located somewhere within a 100-m (328-foot) box, so that the ground station antenna can be pointed to the correct point in the sky. In this case, the use of just three perturbation forces—gravity anomalies, third-body forces, and solar radiation pressure—are quite adequate in achieving this objective.

Finally, before we discuss orbit perturbations, the question arises, why are we interested in this topic? (Readers who are not interested should move on to the next chapter. The rest of the chapters in this book do not require reading this chapter first.) There are two reasons why this topic is of interest. First, an important requirement in the operation of a spacecraft is to know its position in space over time. For example, the operators need to know with confidence that at a specific time tomorrow the spacecraft will rise above the horizon of a ground station, so that it can be commanded to do tasks and it has an opportunity to transmit its payload data to the ground. Clearly, the real perturbed orbit characteristics need to be used in order to make this prediction accurately. Also, in designing low Earth orbit operations, such as the rendezvous of a space shuttle with a space station, orbital perturbations need to be included in the analysis. If they are not, then when you think the two spacecraft have come together in orbit, you will find that they are still tens or even hundreds of kilometers apart! This is especially true if the perturbations due to gravity anomalies in low Earth orbit (LEO) are neglected in the mission planning.

Second, the action of orbit perturbations leads to a requirement for the spacecraft to control its orbit. To understand this, we need to recall the discussion in Chapter 2 about how the best orbit is chosen for a particular spacecraft mission. In summary, the mission orbit is chosen to put the spacecraft into the best place in order that its payload can most effectively achieve the mission's objectives. Having chosen the mission orbit, the spacecraft will then be launched there. Now, if the orbit was ideal, then its shape, size, and orbital inclination would remain fixed, and the spacecraft would continue to reside in its best mission orbit. But we now know that when we launch into the mission orbit, perturbations will change the orbit's attributes over time, which means that the orbit evolves ultimately into one that is inappropriate for the mission objectives. This leads to a requirement to fire small rocket thrusters on the spacecraft to counter the effects of orbit perturbations, so that the spacecraft stays in the mission orbit. This process is referred to as *orbit control,* and is planned and executed routinely by the operators of the spacecraft.

Gravity Anomalies

If the Earth were perfectly spherical, and its internal density distribution had a particularly simple form, then the gravity field of Earth would be a perfect inverse square law, as described by Isaac Newton. However, Earth is not perfectly spherical, nor does it have a simple internal mass distribution. There are topographical features that spoil that perfection, such as mountains that are 8 km (5 miles) high and ocean trenches that are 11 km (7 miles) deep. In addition, Earth's shape is basically that of an oblate spheroid (Fig. 3.1)—a squashed sphere—such that if you were to stand on the poles, you would be 21 km (13 miles) closer to Earth's center than if you stood on the equator.

This is not too surprising given that Earth rotates on its axis once per day, causing the equator to bulge and the poles to flatten. This degree of flattening of Earth—21 km in a radius on the order of 6400 km (4000 miles)—is not too noticeable, however. For example, if you were to stand on the moon and look back at Earth, this degree of oblateness would probably not distract you from the apparent spherical perfection of the brilliant blue orb in the black sky. However, if you think of it in terms of the possible perturbing effects on a spacecraft in a LEO, there is an extra 21 km of crust—of gravitational mass—at the equator, which will have a significant

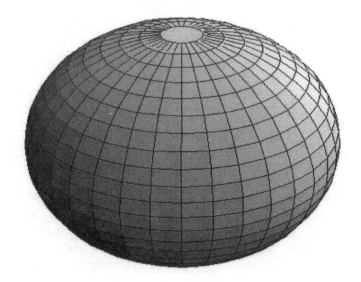

Figure 3.1: The basic shape of Earth is an oblate spheroid, flattened at the poles and bulging at the equator. Note that the actual oblateness of Earth is much less than the shape illustrated!

effect on the spacecraft's motion as it flies over the equatorial region. The gravity field of Earth deviates from that proposed by Newton, causing gravity anomaly orbit perturbations.

What effect does the oblate shape of Earth have on the motion of orbiting satellites? The answer to this is a fairly complex affair, and one that has always challenged me in my career as a teacher in space engineering, even when I am allowed to use mathematics! There are two main effects on the orbit, perigee precession and nodal regression, both of which produce major changes in the orbital motion.

Perigee precession is a gravity anomaly perturbation that affects elliptic orbits. In an ideal elliptic orbit, the major axis of the ellipse, that is, the line from perigee to apogee, remains fixed in direction. If you decided you wanted your spacecraft to be in an orbit with the perigee point above the North Pole, then you would launch into a polar orbit in such a way as to achieve this. In the absence of gravity anomalies, the perigee would remain frozen above the North Pole as required. However, the presence of the extra gravitational mass around the equator, due to Earth's oblateness, causes a greater acceleration on the spacecraft in the perigee region, which in turn causes the trajectory to curve a little more acutely. As a result, when the spacecraft climbs to its apogee, the apogee point has moved, causing the line of the major axis to rotate in the plane of the orbit. This is illustrated in

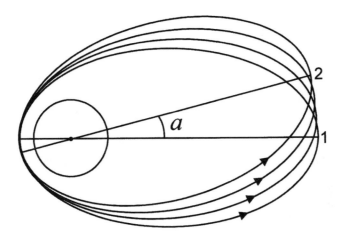

Figure 3.2: Perigee precession causes the major axis of an elliptical orbit to rotate in the plane of the orbit, such that the initial major axis 1 will rotate to major axis 2 after a number of orbit revolutions.

Figure 3.2, where the major axis of the orbit has rotated through an angle a (from 1 to 2) over four orbit revolutions.

In fact to move the major axis through this angle may require many orbit revolutions, particularly for the size of orbit shown, but the diagram has exaggerated the effect to aid clarity. Coming back to our orbit with the perigee above the North Pole, due to the effects of perigee precession, the perigee point will not stay fixed above the Pole but will move steadily around the orbit.

The rate at which the perigee moves is dependent on the size, shape, and orbital inclination of the orbit, so it is difficult to generalize. One thing that can be said, however, is that low orbits are affected more than high ones, since the influence of the extra mass due to the equatorial bulge is more strongly felt when a spacecraft makes a low pass over the equator. To give you an idea of the numbers, for a low elliptical orbit with a perigee height of 300 km (185 miles), an apogee height of 500 km (310 miles) and an orbital inclination of 30 degrees, the major axis of the orbit will rotate at a rate of about 11 degrees per day. To get a feel for altitude dependence, if we stick with the same orbital inclination but increase the perigee altitude to 1000 km (620 miles) and the apogee altitude to 10,000 km (6200 miles), then the perigee precesses at a rate of about 2 degrees per day. In terms of magnitude, this is a large perturbation, even for the higher orbit. Other changes to the orbit that mission analysts get excited about, due to other types of perturbation, are typically of the order of fractions of a degree per day!

Nodal regression is a gravity anomaly perturbation that affects both circular and elliptic orbits. The first question is, What is a node? You may remember the answer from Chapter 2. This is just the point on the orbit where the spacecraft crosses the equator. Clearly it does this twice on each orbit revolution, once when traveling from south to north, and once on the other side of the orbit when traveling from north to south. When the spacecraft motion is "ascending" from south to north, the equator crossing is called the *ascending node,* and when descending from north to south we have a *descending node.* The line between the nodes—the intersection of the orbit plane and the equatorial plane—is called the *line of nodes.* For an ideal orbit, the line of nodes remains fixed in direction with respect to the distant stars, but in a real orbit the gravity anomaly perturbation causes it to move around the equator. This nodal movement—or *nodal regression*—is illustrated in Figure 3.3a, and is due to the extra mass associated with Earth's equatorial bulge.

For orbital inclinations less than 90 degrees, the node moves west around the equator (as shown). When the inclination is 90 degrees—a polar orbit— the node remains stationary, and when the inclination is greater than 90 degrees the node moves east. As the node moves, the orbit plane rotates, while the orbital inclination remains constant. This nodal movement, and plane rotation, will continue indefinitely, as shown in Figure 3.3b. The tip of the arrow describes a circle, while the arrow always remains at right angles to the orbit plane. How nodal regression is produced by Earth's oblate shape is a little difficult to explain, but it is related to something called *gyroscopic precession.*

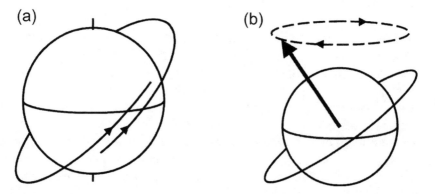

Figure 3.3: (a) In an ideal orbit the node remains stationary, but Earth oblateness causes the node to move so that the spacecraft crosses the equator at a slightly different position on each orbit. (b) Another way of describing the effect is to imagine an arrow that always remains perpendicular to the orbit plane. Over time, Earth's oblateness causes the tip of the arrow to describe a circle.

If you are interested to know what this is about, and perhaps have a bit of a technical background, then read the Nodal Regression box. If not, then you can skip it without compromising your understanding of what follows.

Nodal Regression

The effect of nodal regression on an orbit is similar to the effect of the precession of a gyroscope, when subjected to a torque. A *torque* is a rotational force, like the force you have to apply to remove a bolt from a wheel when you get a flat tire. You apply a force in a rotational sense by pushing at the end of the handle of the wrench. The size of the applied torque is not just to do with the amount of force exerted but also the length of the handle of the wrench you are using. The longer the handle, the greater the "moment arm" and the more torque there is. The axis of the torque is parallel to the direction about which the rotation takes place—in this case the long axis of the bolt.

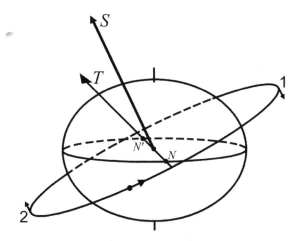

Now, if we refer to the diagram, Earth is depicted as a rather exaggerated oblate sphere. The orbital motion of the spacecraft about Earth produces a spin axis *S*, called an *orbital angular momentum vector,* which is perpendicular to the orbit plane. At the northernmost position on the orbit, point 1, the gravity force on the spacecraft is deflected slightly downward to the extra mass around the equator, producing a small out-of-plane component of gravity as shown. Similarly, at point 2 a small out-of-plane force is produced, but this time directed upward. The combination of these small forces produces a torque on the orbit shown as the arrow *T*, the direction of which is aligned with the orbit's line of nodes *N* − *N'*. As with a

---

gyroscope, if the orbit plane is torqued in this way, its spin axis will tend to align itself with the torque axis; that is, the angular momentum vector S will precess toward the torque vector T. Put more simply, the spin axis S will tilt toward the torque axis T. Since the spin axis S is always perpendicular to the orbit plane, as S tilts so does the orbit plane, causing the node N to move westward along the equator. Also, given that the torque axis T remains parallel to the line of nodes, its direction also rotates westward in the orbit plane. The result is a precessional motion of the orbit spin axis as shown in Figure 3.3b.

How big an effect is nodal regression? If we take a circular orbit typical of a space shuttle, for example 300 km (185 miles) altitude with an orbital inclination of 30 degrees, then the orbit node will move at a rate of about 7 degrees per day in a westward sense. This is again a huge effect compared to other types of perturbation. The effect is less for higher orbits, however, since the spacecraft is further away from the extra mass associated with the equatorial bulge of Earth. For example, a circular orbit at 10,000 km (6200 miles) altitude, with the same inclination, has a modest nodal regression rate of about 0.3 degrees per day.

The bottom line of this discussion about the effects of Earth oblateness is that they are really, really important in low Earth orbit spacecraft operations. If they are neglected, then the position of the spacecraft over time will be in error by many thousands of kilometers!

Gravity Anomaly Perturbations in GEO

Another kind of gravity anomaly perturbation is nicely illustrated by considering the motion of a spacecraft in geostationary Earth orbit (GEO). As we are now aware, the Earth's shape is predominantly that of a squashed sphere, but there is another aspect of the Earth's shape that is surprising. If you were to slice the Earth through the equatorial plane, the shape of the resulting cross section is not circular but approximately elliptical. The Earth's shape can be represented, in an exaggerated way, by the outline shown in Figure 3.4. To get this shape, we take a sphere and give it a good squeeze at the poles to make it oblate, and then give it a small squeeze at the equator, to make the equatorial cross section slightly elliptical. We end up with a form defined by three perpendicular axes, a, b, and c, each having different lengths. From our previous discussions, we know that the equatorial radius b is greater than the polar radius a by about 21 km (13 miles). But now we are suggesting that b and

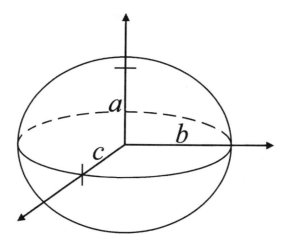

Figure 3.4: The shape of Earth can be approximated by a shape spanned by three axes, all of different length. As well as being oblate, Earth also has an equatorial cross section that is slightly elliptical.

c are different also, but this time by a small amount—less than a kilometer. If we recall that the equatorial radius of Earth $b = 6378$ km (3963 miles), then this difference between b and c means that the degree of ellipticity of the equatorial cross section is small indeed. However, we now focus attention on this, and discuss the effect it has on the motion of a GEO spacecraft.

But how can such a small deviation in the shape of the equatorial cross section have any effect on an orbiting spacecraft? The answer to this question comes from the fact that although the perturbing forces are small, they tend to act on the spacecraft in the same way on each orbit revolution, so that lots of small changes accumulate to cause an effect that is sizable.

To explain this, in Figure 3.5 we are looking down on Earth and the GEO. The elliptical cross-section of the equator is shown in a rather exaggerated way, with the Greenwich Meridian drawn in to relate the ellipse to Earth's geography. In Figure 3.5a, the points A and B represent the "bulges" in the equatorial cross section that occur at around 160 and 350 degrees longitude east, respectively (it may be helpful to have a globe of the world handy while reading this chapter to help with the geography). This places them in the region of the western Pacific Ocean on the one hand, and the west African coast on the other. A spacecraft is also shown in GEO at an arbitrary position, which happens to be at a longitude of about 40 degrees east. Note that the longitudinal position of the GEO spacecraft would depend on what region it serves; in this case it just happens to be stationed above east Africa and the Middle East.

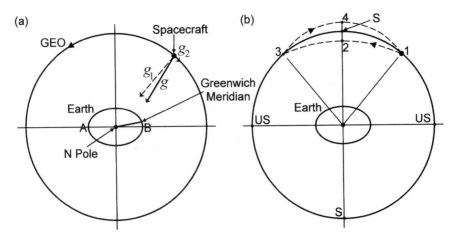

Figure 3.5: (a) Looking down on the elliptical equator of Earth and the GEO. (b) The oscillatory motion of a GEO spacecraft due to the gravity anomaly perturbations caused by the elliptical equator.

An important thing to note about Figure 3.5a is that, for an ideal GEO, the geometry is unchanging. What I mean is that the Earth rotates once per day, in the same time as the spacecraft takes to complete one orbit. The situation is equivalent to rotating the whole of Figure 3.5a about the North Pole axis once per day. In this process the relative positions of Earth and spacecraft do not change. If we now focus on the spacecraft, the gravity force acting on it is modified by the elliptical mass distribution around Earth's equator. Because the spacecraft is closer to the "bulge" at B than that at A, the direction of the gravity force, indicated by the arrow labeled g, is deflected slightly toward B, rather than pointing precisely at Earth's center. The deflection is exaggerated in the diagram for clarity; in reality it is tiny. The resulting force acting on the spacecraft can now be resolved into two *components*, the main one labeled g_1 directed toward Earth's center, and a tiny force component g_2 acting in the local horizontal direction at the spacecraft.

To illustrate the concept of resolving forces into components, a good example is the use of a heavy roller to flatten the bumps in a lawn. If we look at Figure 3.6, there are basically two ways of doing this: we can either push the roller or pull it. Does it make any difference which way we choose? Well, if we resolve the forces into components as shown, we can see that it does. When we push the roller (Fig. 3.6a), the force we use (the solid arrow) can be resolved into two force components (the broken arrows)—one directed horizontally to move the roller along, and another vertical component directed downward. When we pull the roller (Fig. 3.6b), there is the same horizontal component, but now the vertical component is directed upward.

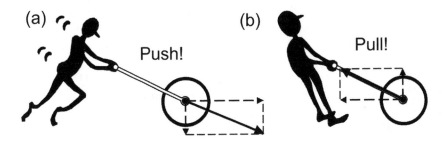

Figure 3.6: The force exerted on a lawn roller (solid arrow) is resolved into horizontal and vertical force components (broken arrows).

If we push the roller, we produce a vertical force component that tends to increase its effective weight and therefore increase friction with the ground, making it harder to move. On the other hand, if we pull the roller, we tend to reduce the ground friction, making it easier to move. The trick of resolving forces into components is often used by engineers, and in the case of the force of gravity acting on a GEO spacecraft, it is just convenient for the argument to resolve it into the vertical and horizontal directions.

Equipped with this intuitive notion of force components, let's return to our discussion about GEO perturbations. Having components of gravity acting in the local horizontal direction is something that we are not very familiar with! In our normal experience, gravity forces always act down the local vertical direction. But in this case the elliptical mass distribution at the equator is producing this rather strange occurrence. The consequences of this for the spacecraft motion are significant. In Figure 3.5b, at point 1 the small horizontal component of gravity force acts in a direction that is opposed to the spacecraft's motion around the GEO; this direction is referred to as *retrograde*. This small retrograde force causes a decrease in the energy of the orbit, with a corresponding decrease in orbit height. What happens if the height of a GEO decreases? The spacecraft orbit speed increases, and it goes round the orbit in slightly less than 1 day; it is no longer synchronous with Earth's rotation. As well as losing height, it also drifts away from its on-station longitude. This situation is depicted by the broken curve from point 1 to point 2. It should be noted that the height change illustrated at point 2 is again exaggerated to make it clear.

At point 2, the spacecraft is at the same distance from the bulges at A and B, and all the gravity force is now directed toward Earth's center. However, because the orbit height is still low, the spacecraft continues to drift, and once it passes point 2 the small horizontal gravity force recurs. But now its direction is reversed, as the spacecraft is now closer to bulge A. The small horizontal gravity force is now in the prograde direction—in the same direction as the spacecraft's motion. This tends to increase the orbit energy, causing the height to increase again, as indicated by the broken curve from point 2 to point 3. At point 3, the spacecraft has regained GEO height, and so becomes synchronous again, halting the drift in longitude. However, the force continues to act in a prograde direction, causing the orbit height to increase above the GEO. Now the orbit speed becomes less than the GEO speed, and the synchronism is again lost as the spacecraft drifts in the opposite direction, represented by the broken curve from point 3 to point 4. Beyond point 4, the force reverses, acting in a retrograde direction once again. This causes a reduction in orbit height until the spacecraft finally returns to its on-station position at point 1.

This rather interesting circuitous journey takes a typical spacecraft quite a long time—of the order of hundreds of days—and is generally a bit of a nuisance for the spacecraft operators, who would prefer the spacecraft to stay at the required on-station position! The operators have to plan and execute orbit control maneuvers to combat the effects of these gravity anomaly perturbations. This means firing small rocket thrusters on the spacecraft (see Chapter 9) to ensure that the spacecraft stays in position. Without this, the spacecraft would oscillate indefinitely in longitude about the stable point S, which is where the minor axis of the equatorial ellipse cuts the GEO arc. In the example above, this would mean that an uncontrolled spacecraft would wander off from its on-station position at 40 degrees to around 120 degrees longitude east (East Africa to Indonesia), and back again on a regular basis. I always find it amazing how such a small variation in Earth's shape can cause such a large change in the spacecraft's orbit!

Note that the orbital positions above the bulges in the equator are unstable—labeled US in Figure 3.5b. Uncontrolled spacecraft positioned near these would move off toward the nearest stable point, labeled S.

Third-Body Forces

Third-body force perturbations are caused by the gravitational influence of a third body in addition to the spacecraft and the Earth. The Earth is not isolated in the universe; there are other celestial bodies out there, the gravitational fields of which can have a significant effect on the motion of an Earth-orbiting spacecraft. The Sun and the Moon have the greatest perturbing effect. If we think about these bodies, and look at Figure 3.7, then the total gravity force governing the motion of the spacecraft becomes a

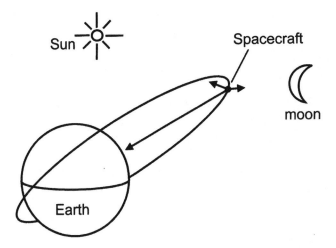

Figure 3.7: The motion of the orbiting spacecraft is influenced not only by Earth's gravity but also by the gravity fields of third bodies, principally the Sun and the moon.

sum of the gravity forces due to the Earth, the Sun, and the Moon. In most applications, the analysis of third-body perturbations includes only the influence of the Sun and the Moon, and then the effects are often called *luni-solar perturbations*. Of course, if high positioning accuracy is required for a particular spacecraft mission, then other third bodies, such as Mars, Venus, and Jupiter, can be included in the analysis until the required degree of precision is achieved.

The effects of third-body forces on low-altitude circular orbits are small. In this case, the Earth's gravity field dominates the contributions from other celestial bodies due to the spacecraft's closeness to the Earth. However, if the spacecraft's orbit takes it to a significant distance from the Earth, for example, in geostationary Earth orbit or at the apogee of a highly eccentric orbit, then the gravitational influence of third bodies is more important. The Earth's gravity force on the spacecraft is reduced because of the greater distance from the Earth's center, whereas the gravitational force due to, say, the Sun has not changed significantly. The overall effect on the orbit remains small, but the ratio of the Sun's gravity force to that of Earth has increased. Therefore, the effects of third-body perturbations are greater in high-altitude orbits. The next obvious question is, What changes do third-body forces produce in the spacecraft's orbit?

In answering this question, an important thing to note is that the gravity field of third bodies generally produce perturbing forces that are out of the plane of the spacecraft's orbit. This is because bodies like the Sun and the

Moon are not often to be found within the plane of the spacecraft's orbit. As a consequence, the main effect of the third-body perturbations is to cause small changes in the plane of the orbit, that is, changes in the orbital inclination. In addition, small oscillations in the size and shape of the orbit are also produced, which can be important for a highly eccentric orbit with a low perigee altitude. In this case, the size and shape variations result in an oscillation in the perigee height, and this can cause the perigee to dip in and out of Earth's atmosphere (see next section on spacecraft aerodynamics). A reentry of the spacecraft into the atmosphere may then result, with the unpleasant prospect of a premature end to the spacecraft's mission life.

Aerodynamic Forces

It seems strange to be talking about aerodynamic effects on spacecraft, because space is a vacuum. Right? Well, as far as people are concerned, as living and breathing creatures, space is *effectively* a vacuum; if you stepped out of a space station in orbit and took off your helmet, then the consequences would reinforce this notion. However, for an orbiting spacecraft the effects of air drag are encountered at heights up to around 1000 km. At these altitudes, there is an atmosphere, but it is extremely thin. To describe how thin it is, let's think about the atmospheric density at an altitude of, say, 800 km (500 miles) at mean solar activity (for a discussion on the effects of solar activity on atmospheric density, see Chapter 6). In every cubic meter of volume at this height there is a mass of air of around 0.000 000 000 000 01 kg, which explains why you can't breath it! Compare this with the density of air at sea level, which is around 1.2 kg/m^3.

The next question is, How can such a thin atmosphere produce aerodynamic effects on spacecraft that perturb their orbital motion (see also the discussion about drag forces on launchers in Chapter 5)? The key to this is to realize that the aerodynamic force on an object is dependent not only on the air density but also on how fast the object is moving through the air. For example, we know that sufficient aerodynamic force can be exerted on a garden fence to knock it down in a winter storm, provided the wind speed is high enough. This force, known as *dynamic pressure*, actually depends on the square of the wind speed. If the wind speed doubles, the force on the fence increases by a factor of four (2^2), if it trebles the force is nine times as big (3^2), and so on. No wonder storm-force winds make short work of fences!

Taking the argument a little higher than a garden fence, we can think about airplanes moving about in the atmosphere. They encounter high-speed winds, but this time the wind is produced by their own motion through the air. The flight of an airplane through the air is resisted by an aerodynamic drag force. This force is measured in a number of different units, but the most common is the Newton, named after Isaac Newton. A Newton of force has a formal definition: it is the force required to accelerate a 1 kg mass by 1 m/sec^2. As we explained in Chapter 1, this is an increase in speed of 1 m/sec for every second that the force is applied. Another perhaps easier way to get a feeling for a Newton of force is to adopt an approximate and informal definition, which is that it is about the weight of a (smallish) apple! Returning to our airplane, and thinking about a civil airliner at cruising altitude, the level of dynamic pressure acting on it due to its motion through the air is on the order of 10,000 to 15,000 Newtons for every square meter of area presented to the air flow. A metric tonne weight is about equal to about 10,000 Newtons, so that's quite a lot of aerodynamic drag, which of course needs to be overcome by the thrust from the jet engines to keep the airplane in the air.

Finally, climbing even higher to orbital altitude, the same principle applies to spacecraft. They encounter a level of aerodynamic drag force that is much smaller than that of an airplane, but it is nevertheless tangible because, although the air density at orbit height is small, the speed of the spacecraft through the atmosphere is high. For each square meter of spacecraft area presented to the air flow, the level of aerodynamic force varies from about 1/100th of a Newton at a 200-km altitude to a tiny 0.000 000 05 of a Newton at a 1000-km altitude. These small forces are produced by the molecules of atmosphere impacting on the spacecraft. The forces seem too small to be of any consequence, but the point is that they act all the time, in a retrograde direction, that is, in a direction opposed to the motion of the spacecraft, producing a small but steady decrease in the orbit height day after day. For example, if the spacecraft resides in a 200-km-altitude circular orbit, this steady decay will lead to the spacecraft's reentering the atmosphere within a short period of time—a few days to a few weeks depending on the characteristics of the vehicle.

We can now begin to understand how aerodynamic drag perturbations change a typical satellite's orbit. The main effects are to reduce the size of the orbit and to change the orbit's shape, making it more circular. To see this, we imagine a spacecraft in an eccentric orbit, as shown in Figure 3.8a, with an initial apogee at point 1, and a perigee height low enough to allow the spacecraft to dip into Earth's thin upper atmosphere. During each perigee passage, the aerodynamic drag forces take energy out of the orbit, so that it does not quite reach the same height on each subsequent apogee (points 2, 3, and so on). In the figure we see two things happening: the size of the orbit is

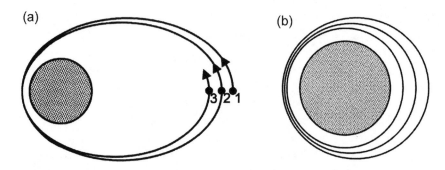

Figure 3.8: (a) The drag effects occur around the perigee of the orbit, mainly causing a change in apogee height. As a consequence, both the orbit size and eccentricity decrease. (b) The evolution of a moderate eccentricity orbit due to air drag, starting at the outer ellipse and ending at the inner circular orbit.

decreasing, and it is becoming more circular. It should be noted that the figure is not drawn to scale, and the sort of decrease in apogee height illustrated would not occur in two orbit revolutions, but in fact would require perhaps hundreds of orbits revolutions. Figure 3.8b shows a similar orbit evolution for an orbit of moderate eccentricity. Again, the major effect is a decrease in apogee height, with a smaller lowering of perigee altitude.

For a spacecraft in a circular orbit, the drag force causes the orbit height to decrease on each orbit revolution, producing a kind of spiral trajectory of diminishing altitude. By how much the orbit height changes depends on the characteristics of the spacecraft. For example, if it is small and massive like the cannonball shot from Newton's cannon in Chapter 2, then the drag effects are relatively small. On the other hand, if the spacecraft is large and of low mass, like a balloon satellite, then the drag effect on orbit height is large. A number of such balloon satellites were launched in the 1960s to bounce radio waves off in early space communications experiments, but their orbit lifetime was generally short due to their vulnerability to drag perturbations. In "ballpark" numbers, for a typical spacecraft the height change per orbit at a 800-km altitude may be on the order of a few centimeters or a few tens of centimeters, whereas at 200-km altitude the height can change by as much as a kilometer or more for each orbit revolution. Clearly, spacecraft in low-altitude circular orbits do not stay in orbit for long.

Another curious feature about drag force on a spacecraft in a low circular orbit is that, although the force acts in the direction opposed to its motion, it actually causes the spacecraft's speed to increase. This is something we do not experience often, and it seems quite counterintuitive. However, a little thought resolves the puzzle. Any force acting in a retrograde sense on a spacecraft will take energy out of the orbit and cause the orbit height to decrease. And as we saw in Chapter 2, the lower the orbit, the faster the spacecraft moves. What's happening here is that, given the decrease in height on each orbit revolution, the spacecraft is actually flying "downhill." The situation is entirely analogous to riding a bicycle down a slope. Gravity pulls it forward and tends to increase its speed, but at the same time we can feel the wind on our faces, which produces a drag force that acts against the motion, tending to slow us down. In the case of the satellite, the gravity pulling it forward is larger than the drag force slowing it down, resulting in the net increase in orbital speed as its orbit height decreases. Surprisingly, I have seen statements in professional journals along the lines of "the aerodynamic drag force decreases the orbit velocity," so you can see that sometimes even the experts get it wrong!

Solar Radiation Pressure

The final orbit perturbation on our brief list of the most important effects causing changes to orbits is solar radiation pressure (SRP). Like drag, the change is produced by a pressure acting on the spacecraft, but this time the pressure is produced by light, in particular the bright sunshine illuminating the spacecraft. Twentieth century physicists were a clever lot, and they first worked out that light reflecting from a surface exerts a pressure on it. As you read this book, the source of light you are using is producing a small force on the page. The fact that nobody noticed this before the 20th century, suggests that light pressure is tiny, and this is indeed the case. You may recall that the drag force is generated by the impact of air molecules on the spacecraft as it speeds through the atmosphere. SRP shares the same mechanism, but the atmospheric particles are replaced in this case by the stream of particles of light—referred to as *photons*—emanating from the Sun.

The magnitude of the pressure is unimpressive, amounting to a few millionth of a Newton for every square meter of spacecraft area presented to the Sun. Comparing this with aerodynamic drag, we find that the magnitudes of drag and SRP effects are about the same for a spacecraft in a circular orbit at around a 600- to 700-km (370- to 435-mile) altitude (depending on the level of solar activity; see Chapter 6). There are differences, however; the magnitude of SRP decreases with the distance from the Sun, as opposed to drag, which decreases with increased height above the Earth. Also, the force of SRP generally pushes the spacecraft away from the Sun, whereas the drag force always acts in a retrograde direction with respect to the spacecraft's forward motion.

It is difficult to summarize the effects that SRP have upon the spacecraft orbit in any meaningful way. Below the 600- to 700-km altitude mentioned above, the SRP perturbations are completely swamped by aerodynamic drag. Above this height, the changes they produce are greatly dependent on the aspect that the orbit plane presents to the Sun. The other thing to remember is that the force is tiny. Generally, small cyclic variations in the size, shape, and orbital inclination are produced. But, as we saw with drag, a tiny force acting on the spacecraft in the same way on each orbit over time can accumulate significant orbit changes. Furthermore, if the spacecraft presents a large area to the Sun—for example, solar panels to convert sunlight into usable electrical power—then the perturbing effects on the orbit are further amplified.

One case where the perturbing effects of SRP can be seen to build up, and explained in a fairly intuitive manner, is that of the motion of a satellite in a geostationary Earth orbit (GEO). This is shown as the circle drawn with a continuous line in Figure 3.9a, seen from the perspective of someone looking down from above the Earth's North Pole.

When the spacecraft is at point 1 in the GEO, the SRP force acts in the direction opposing the motion, causing a small decrease in orbital energy. As a consequence, the orbit height achieved on the opposite side of Earth is reduced, thus forming a perigee at point 2. At this point, the SRP force pushes the spacecraft along, tending to produce a small increase in energy that takes the spacecraft to a higher altitude at point 3. The combination of these effects transforms the circular GEO into the elliptic orbit illustrated in Figure 3.9b, which has its major axis aligned at right angles to the direction of the sunlight. As usual, the discussion has been somewhat simplified; for example, the typical eccentricity produced by SRP perturbations in a GEO is generally much less than that shown in Figure 3.9b. Also, it takes many orbit revolutions for the perturbation to build up this moderate eccentricity, rather than the one revolution discussed above. But the message is clear: SRP perturbations increase the eccentricity of a GEO from an initially circular state to an elliptical one. Why is this important?

If you recall the discussion of the GEO orbit in Chapter 2, its main advantage is that a spacecraft in GEO remains stationary with respect to a ground-based observer, so that communications dishes do not have to move to maintain a link. But this is only true if the orbit is circular, when the spacecraft's speed remains constant. If the orbit becomes slightly elliptical, due to the effects of SRP, then the spacecraft moves a little faster than circular orbit speed at the perigee point of the orbit, and a little slower at the apogee point. From the perspective of someone at the ground station, the spacecraft no longer remains stationary at the point where the commu-

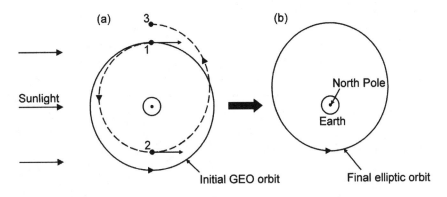

Figure 3.9: The solar radiation pressure perturbation produces an eccentricity in an initially circular orbit.

nications dish is directed, but it appears to wander back and forth through this point, with a period of 24 hours. To counter this, the spacecraft again needs to perform orbit control maneuvers to prevent the SRP perturbations building up. The operators of the spacecraft command it to fire small rocket engines (thrusters) to keep the orbit circular. Although light pressure does seem a strange idea, its reality is confirmed by the fact that spacecraft engineers have to estimate an amount of thruster fuel to put in the spacecraft's tanks in order to control its effect on the spacecraft.

Summary

As can be seen from the above discussion, the topic of orbit perturbations is a fairly complex one. It has been my aim to demonstrate that perturbations need to be taken into account when doing the mission analysis for a real spacecraft project. Another aspect that is implicit in the discussion is that dealing with the perturbation effects requires a good deal of mathematical and computational expertise, which is a routine part of any spacecraft mission analysis activity.

The differences between the ideal, Keplerian orbits of Chapter 2, and the real orbits discussed in this chapter are nicely summed up by the geostationary Earth orbit (GEO). For example, if we have a communications satellite in GEO, then in the absence of perturbations we simply launch the spacecraft into a circular, equatorial orbit at the right height to give an orbit period of one day (see Chapter 2). To someone on the ground, the spacecraft then appears to remain stationary in the sky, and the various communications dishes on the ground that wish to use the satellite simply stare at this fixed position. However, the situation is a little more arduous for the satellite operators in the real world when the effects of perturbations have to be countered. As we have seen, there are three main perturbations that affect a GEO satellite: gravity anomalies, luni-solar perturbations, and solar radiation pressure. Each of these has a distinctive signature with respect to the motion of the satellite as seen from the ground. Gravity anomalies cause the satellite to move away slowly in an east–west (or longitudinal) direction from the point in the sky to which the ground station dish is directed, and in some cases this can cause the spacecraft to disappear over the horizon! Luni-solar perturbations cause changes in the orbital inclination, which in turn cause the satellite's position to oscillate in a north–south (or latitudinal) sense with a period of 1 day. Finally, we have seen that solar radiation pressure effects produce an eccentricity in the orbit, which leads to an east–west (or longitudinal) oscillation as well. Keeping the spacecraft in the line of sight of the ground dish is a nontrivial orbit-control exercise.

The rest of the chapters in this book are less technically challenging than this chapter. With this basic background in orbits, we now move on to Chapter 4 to look at some mission orbits that are a little more exotic than the popular operational orbits that we have already looked at.

Beyond Circles and Ellipses

THE first three chapters discussed orbits and the work of space mission analysts. However, there are a variety of complex space missions these days that use trajectories that do not conform to the predominantly circular or elliptical orbits that we have discussed so far. Some of these, such as hyperbolic swing-bys, could be described as *ideal trajectories* in the sense that they remain under the umbrella of Isaac Newton's theory. However, recalling the hyperbolas briefly discussed in Chapter 1, these trajectories represent open trajectories that in no way resemble circular or elliptic orbits. In addition, some spacecraft operate in places where the gravitational (and other) forces acting upon them do not approximate to Newton's inverse square law. In these cases the spacecraft's orbit may not resemble a circle or an ellipse at all, and can be described as completely *non-Keplerian*.

In this chapter, we look briefly at some of these unusual trajectories that are becoming more popularly used by spacecraft operators. In particular we look at swing-by maneuvers past planets, orbits around small irregularly shaped bodies such as asteroids or comets, and halo orbits around Lagrangian points.

Swing-By Trajectories

If a spacecraft passes close by a planet, as it journeys through interplanetary space, then the path it takes is described as a *swing-by trajectory*. It may be worth recalling some of the background we discussed in Chapter 1 about this type of trajectory, and indeed you may wish to reread the text associated with Figures 1.9 and 1.10 to refresh your memory. The shape of the curve describing the spacecraft's path is called a hyperbola, and it is one of the four basic *conic section* shapes found by Isaac Newton in his equations describing motion in an *inverse square law gravity field*. As mentioned in Chapter 1, the shape can be seen all over the place once you start looking for it. You may even have your very own hyperbola in the room where you are reading this,

G. Swinerd, *How Spacecraft Fly: Spaceflight Without Formulae,*
DOI: 10.1007/978-0-387-76572-3_4, © Praxis Publishing, Ltd. 2008

if you have a table or standard lamp placed adjacent to a wall (see Figure 1.10).

The shape of the hyperbolic swing-by trajectory is shown in Figure 4.1. When the spacecraft is a long way away from the planet, before it even gets to point A in the diagram, the gravity force of the planet is so feeble that the spacecraft effectively travels in a straight line. This line is called an *asymptote* of the hyperbola. However, as the spacecraft closes in on the planet, the gravity force steadily increases, and the spacecraft's path describes the classic hyperbolic shape depicted in Figure 4.1 from point A through point B to point C. Beyond point C, the gravity force decreases rapidly, and the trajectory tends to the straight line given by the asymptote once again. During this process, the planet has changed the direction of travel of the spacecraft, and this change is given by the *deflection angle,* which is effectively the angle between the incoming and outgoing asymptotes, as shown. The amount by which the trajectory is deflected is

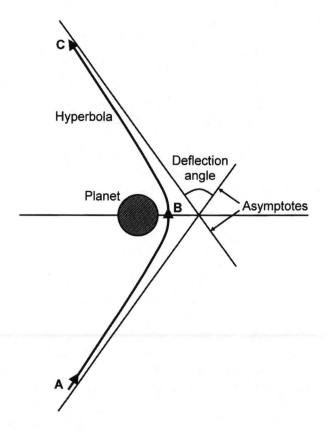

Figure 4.1: The classic hyperbolic shape of a swing-by maneuver past a planet.

dependent on three things: how massive the planet is, how close point B is to the planet, and how fast the spacecraft is traveling on approach.

This description echoes much of what was said in Chapter 1, but one thing that was not discussed was the speed of the spacecraft on the hyperbola. As the spacecraft falls toward the planet on the hyperbola, the gravity force increases, causing a corresponding increase in the vehicle's speed. Conversely, after the point of closest approach, the speed decreases again as the spacecraft climbs away against the force of gravity. The important thing to remember for what follows is that as speed and height are traded, and no energy is lost in the encounter, the speed of approach at A is identical to the speed of departure at C, relative to the planet. This is analogous to a cyclist negotiating a road that falls into a valley between two hilltops. The cyclist leaves the first hill at, say, 10 mph, and accelerates on the downhill section, reaching a maximum speed at the bottom of the valley. The cycle then slows down steadily on the uphill gradient on the other side of the valley, finally reaching the second hilltop at 10 mph again. To make the comparison with the spacecraft's motion complete, we have to assume that no energy is lost by the cyclist due to things like friction with the road or wind resistance. But nevertheless it is quite a good way of thinking about how the speed of the spacecraft varies on the hyperbola.

Gravity Assists

However, this is not the end of the story of hyperbolic swing-bys, because swing-bys can be used to increase the speed of spacecraft in their travels around the Sun without firing a rocket motor. (Swing-bys can also be used to decrease spacecraft speed, but we will not discuss this aspect.) Mission analysts get excited at the prospect of gaining spacecraft energy without having to use rocket fuel! It means that their spacecraft can achieve the journey to a distant planet more quickly, and do it without having to exchange payload mass for propellant mass. This type of maneuver is often referred to as a gravity assist, and it has been used many times by mission designers.

Perhaps the most famous example of its use is in the Voyager spacecraft program to explore the outer solar system. If we look back on the history of astronautics in the 20th century, the events that stand out for me are Sputnik 1, the first manned orbital flight by Yuri Gagarin, the Apollo moon landings, and the exploration of the outer solar system by the Voyager 1 and 2 spacecraft. Voyager 2 was launched in 1977, and took advantage of a rare alignment of the planets to visit Jupiter, Saturn, Uranus, and Neptune before finally leaving the solar system. The scientific return was huge! And it was achievable through the use of gravity-assist maneuvers at each planetary encounter, which increased the spacecraft's speed relative to the Sun so that the Neptune fly-by could be achieved after just 12 years from launch. Without this technique, and using the same launch energy, the transfer to Neptune would have taken around 30 years, and we would only now be learning about Neptune's mysterious moon Triton as I write this in 2006 (assuming that the spacecraft would still be operating after 30 years!).

Gravity-assist maneuvers clearly have great benefit, but how exactly do they work? Let's consider gravity-assist maneuvers using the planet Jupiter as an example. Given that Jupiter is the largest planet in the solar system, it has the strongest gravity field, and so the effect it has on the dynamics of a passing spacecraft is very significant. Let's imagine our spacecraft wandering through interplanetary space, traveling out from Earth toward Jupiter. We also suppose that, in this interplanetary cruise, the spacecraft is in an elliptical orbit around the Sun, so that it is the Sun's gravity that governs its motion. However, when the spacecraft is about 40 million kilometers (25 million miles) from Jupiter, the gravity field of Jupiter begins to be about the same as that of the Sun. Beyond this point, as the spacecraft closes in on Jupiter, the force of Jupiter's gravity increases, and once it is within, say, 10 million kilometers (6 million miles) of Jupiter, the Sun's influence is negligible. At this position, point A in Figure 4.2, the spacecraft is effectively falling toward Jupiter, moving in a (more or less) straight line with a speed given by V_{in} relative to Jupiter. In this part of the flight, while governed by Jupiter's gravity field, the spacecraft performs the classic hyperbolic trajectory as described above. It reaches a maximum speed at closest approach (point B), and then climbs away arriving at point C traveling effectively in a straight line once again with a speed V_{out} relative to Jupiter. From our discussion above about the spacecraft's speed on the hyperbola, we know that V_{in} and V_{out} are the same, so we don't seem to have gained anything from the maneuver! Remember, however, that these are *speeds relative to Jupiter*. The thing we have forgotten is that Jupiter itself is moving along its orbit around the Sun at about 13 km/sec (8 miles/sec) relative to the Sun, and this makes all the difference.

In Figure 4.2, the speeds of objects are represented by arrows. The orientation of the arrow represents the direction in which the object is moving, and the length of the arrow indicates how fast it is going: long arrows denote fast objects, and short arrows denote slow objects. These arrows are called *velocity vectors*, which are useful tools in the analysis of this type of problem. Incidentally, in Chapter 3 we used force vectors, although we did not describe them as such, in the discussion of orbit perturbations, where the direction of an arrow indicated the direction of a force, and the length of the arrow indicated how much force was applied. There are lots of

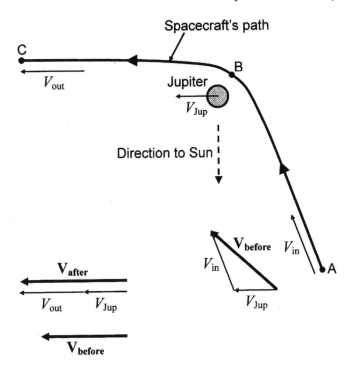

Figure 4.2: A schematic of a Jupiter gravity-assist maneuver.

objects in physics and dynamics that require two pieces of information—direction and magnitude—to describe them fully, and these are all described as vectors by scientists and engineers. For example, there are position vectors, velocity vectors, force vectors, torque vectors, and others, but I'm getting away from the point here.

Getting back to our gravity-assist maneuver, what defines the orbit around the Sun before the swing-by is the position and velocity of the spacecraft relative to the Sun before it enters Jupiter's gravitational influence. Let's suppose that point A in Figure 4.2 is now right at the edge of Jupiter's sphere of influence, approximately 40 million kilometers (25 million miles) out from the planet. If V_{in} again represents the incoming velocity of the spacecraft relative to Jupiter at A, and V_{Jup} is the velocity of Jupiter on its orbit around the Sun, then we can calculate the velocity of the spacecraft relative to the Sun before the encounter. This is denoted by $\mathbf{V}_{before}$ in the figure, and it is the sum of the velocity of the spacecraft relative to Jupiter and the velocity of Jupiter relative to the Sun. Given that the arrows, or vectors, representing these velocities are not parallel, we have to use a *velocity vector diagram* to do this sum, and this is shown as the triangle in the lower right of the figure. The important thing to note is that $\mathbf{V}_{before}$

represents the velocity of the spacecraft on its elliptical orbit around the Sun before meeting Jupiter.

Working out the corresponding velocity of the spacecraft relative to the Sun afterward is a little easier, as in this case the velocities V_{out} and V_{Jup} are parallel to each other, so $\mathbf{V_{after}}$ is just a simple arithmetical sum. This is shown in the lower left in Figure 4.2, opposite the vector triangle. V_{out} and V_{Jup} need not necessarily be parallel to each other, but I have devised the gravity-assist geometry so that they are, to make things a little easier to visualize. Remember that $\mathbf{V_{after}}$ represents the spacecraft's velocity relative to the Sun after the gravity assist, and so defines the subsequent orbit around the Sun after the encounter with Jupiter. In the bottom left of Figure 4.2, the magnitudes of the Sun-relative speeds before and after the encounter—$\mathbf{V_{before}}$ and $\mathbf{V_{after}}$—are compared, and it is easy to see that the spacecraft has gained speed.

The next question to ask is, Where has that speed gain come from? The answer is that the spacecraft's gain is Jupiter's loss. The planet has tugged on the spacecraft, to give it a significant boost in speed, while at the same time the spacecraft has exerted an opposite tug on Jupiter, causing it to lose speed. However, given the mass of the spacecraft compared to the huge mass of Jupiter, the effect on Jupiter is immeasurably small, although it is measurable if you wait long enough; in the words of a National Aeronautics and Space Administration (NASA) press release about the Voyager Jupiter swing-by, "The position of Jupiter will change by about 1 foot every trillion years"!

Figure 4.3 shows a remarkable plot of the Sun-relative speed of Voyager 2 during its interplanetary cruise to the outer planets. The "blips" in the curve at around 5, 10, 20 and 30 astronomical units (AU) correspond to the gravity-assist maneuvers at Jupiter, Saturn, Uranus, and Neptune, respectively. You may recall from Chapter 1 that 1 AU is equal to the mean Earth–Sun distance. These maneuvers maintained the speed of the spacecraft well above that needed to escape the Sun, which is indicated by the broken line.

We can get a feel for the way gravity-assist maneuvers work by proposing a useful analogy, in the form of a rather unusual experiment involving a double-decker bus. Firstly, I would definitely recommend that you do not attempt to perform the experiment, as I would not wish to be responsible for any resulting injuries—and secondly, it allows the rather unusual notion of having a picture (see Figure 4.4a) of a double-decker bus in a book about spaceflight! Those of you who are familiar with the old-style double-decker bus know that the entrance is via a step-up, open platform at the back of the bus as shown in Figure 4.4, with a vertical handrail for passengers to hang onto to prevent them from falling out while the bus is moving. The vertical handrail plays the role of the planet Jupiter in the experiment!

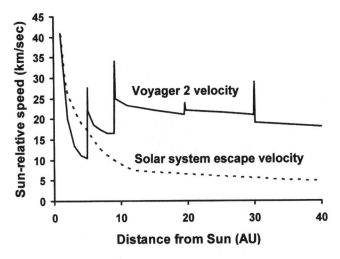

Figure 4.3: The speed of the Voyager 2 spacecraft relative to the Sun during its 12-year mission to the outer planets. (Figure compiled from data courtesy of Steve Matousek, National Aeronautics and Space Administration [NASA]/Jet Propulsion Laboratory [JPL]—Caltech.)

The layout of the experiment is illustrated in Figure 4.4b, which shows a view of the bus looking down from above. The vertical handrail is positioned at the rear corner of the bus, where the open platform entrance is located. Note that the entrance has been switched to the right-hand side of the bus, to allow us to compare the geometry more easily with Figure 4.2. A person, shown rather unimaginatively as a blob with an extended arm and hand, is shown standing on the road, swinging on the handrail. Surprisingly, this rather unlikely arrangement gives a good depiction of what happens in a gravity-assist maneuver, as discussed above. Each of the items shown in Figure 4.4b represents some part of the real thing. For example, as mentioned above, the vertical handrail represents the planet Jupiter, where its movement over the ground, as the bus moves forward, represents the movement of the planet in its orbit around the Sun. The person swinging on the rail represents the spacecraft, and the force in the person's arm as he swings on the rail plays the role of the gravitational attraction between the planet and the spacecraft. Now we can play the experiment two ways—first with the bus stationary and then with it moving.

The first task for our willing helper is fairly simple then, as we would just like them to run up to the rear of the stationary bus at, say, 10 mph (4.5 m/sec)—see plan view in Figure 4.5—grab hold of the vertical hand rail at point 1, and then release his grip at point 2. This simply has the effect of changing the direction of his run from the vector V_{in} to the vector V_{out} as

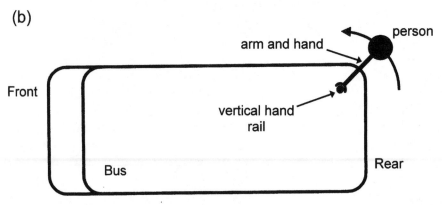

Figure 4.4: (a) The back of a bus! Note the vertical handrail in the open, step-up platform entrance. (b) Plan view of bus, showing relevant features of the thought experiment described in text.

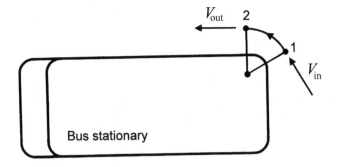

Figure 4.5: Depiction of a hyperbolic swing-by, using the bus analogy.

shown, without effectively changing his speed. This parallels a hyperbolic swing-by trajectory, where the incoming and outgoing speeds relative to the planet are the same but the direction of travel of the spacecraft is changed.

To go any further with the experiment, we have to suppose that our helper is fairly athletic, and has good hand–eye coordination, as we will now expect them to grab the handrail while the bus is moving. Not wishing to make this too difficult, we'll constrain the speed of the bus to 10 mph. This situation is now shown in plan view in Figure 4.6. Again, the runner grabs the rail at point 1 and releases it at point 2. To make a proper parallel with the

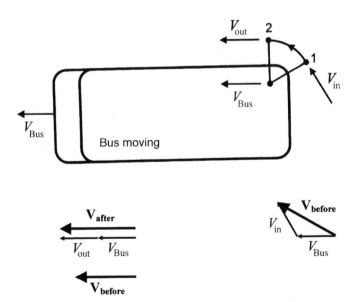

Figure 4.6: Depiction of a gravity-assist maneuver. The speed of the runner over the ground is increased by the movement of the bus. Compare with Figure 4.2.

spacecraft gravity assist, the runner has to perform his swinging movement on the rail in such a way that it looks the same as the stationary bus swing from the perspective of someone standing on the moving bus platform. But then, perhaps this is one technical detail too far? The important thing to take away from the moving bus analogy is an intuitive feel that once the runner grabs the rail, he acquires additional speed over the ground from the movement of the bus. If you imagine grabbing the moving handrail yourself, you can almost feel the stress in your arm, tending to pull the socket, as your body mass is accelerated by the bus! The force in your arm will also act on the bus in the opposite sense, tending to slow it down a little, in the same way as we saw Voyager taking a tiny amount of orbital speed from Jupiter. The speeds over the ground of the runner V_{before} and V_{after} are worked out at the bottom of Figure 4.6, in an analogous way to that done in the swing-by case in Figure 4.2, confirming an increase in speed. Note that the speed over the ground in this analogy corresponds to the speed of a spacecraft relative to the Sun in the gravity-assist maneuver.

This analogy presents a situation in which you can at least imagine experiencing a boost in speed in a way that is similar to what happens in a gravity assist.

Orbits Around Small, Irregularly Shaped Bodies

Recently there has been a lot of interest in spacecraft missions that visit the smaller objects in the solar system, such as asteroids and comets. These types of objects are essentially debris scattered across interplanetary space, which are fragments left over from the processes that formed the Sun and planets. And therein lies their attraction as targets of scientific interest for spacecraft probes.

Asteroids, sometimes called minor planets, are usually solid bodies which vary in size from around 900 km (560 miles) in diameter to tiny boulders a few meters across. The majority of these objects travel in orbits between Mars and Jupiter, although many of the smaller objects can be found almost anywhere in the inner solar system. Comets are also small objects, typically about 1 to 10 km in diameter, composed mostly of ice and dust. The current view is that these "dirty snowballs" originate from a region of space distant from the Sun, called the *Oort cloud,* around tens of thousands of AU away from the Sun. Periodically, a comet will be knocked out of the cloud by the gravitational disturbances of a passing star, causing it to fall into the inner solar system. As it does so, once it is closer than about 5 AU from the Sun, the ice begins to react to solar heating, causing plumes of gas and dust. This produces the characteristic appearance of comets—a compact nucleus and a long tail, shining magnificently by reflected sunlight. In past times, they were often greeted with a mixture of awe and suspicion, and were sometimes regarded as an omen of some impending disaster. Fortunately, the public is now better informed, although a strong reaction can still be evoked by a bright comet.

These small bodies are believed to contain crucial clues to aid understanding of the origin and early evolution of our solar system, which is why there has been growing interest in sending spacecraft to investigate them. The first orbital mission around an asteroid was achieved by the Near Earth Asteroid Rendezvous (NEAR) Shoemaker spacecraft, which entered orbit around its target asteroid Eros in February 2000. The spacecraft was not designed to land on the asteroid's surface, but the mission was finished off in February 2001 with a rather ad hoc descent and touchdown on Eros's surface, making for a first in astronautical history. The other high-profile mission in progress at the time of this writing is the European Space Agency's Rosetta mission to orbit and land on a comet. Rosetta's rendezvous with its intended target will not occur until May 2014. On arrival, the Rosetta spacecraft will enter orbit around the small comet 67P/Churyumov-Gerasimenko, finally deploying a small instrument package to make a soft landing on the comet's nucleus.

If we think about orbits around asteroids and comets, they do not fall into the ideal-orbit category discussed in Chapters 2 and 3. You may recall that a so-called ideal orbit is one that results from the motion of a spacecraft in a pure inverse square law gravity field. This produces the familiar trajectory shapes called conic sections: the circle, the ellipse, the parabola, and the hyperbola. The main reason why orbits around asteroids and comets do not fall into this category is that these objects are usually irregular in shape, so that their gravitational fields are nothing like that described by Newton's inverse square law. This affects the orbital motion significantly.

You may also remember from Chapter 2 that the motion of a spacecraft in an ideal circular or elliptical orbit takes place in a fixed plane, and is periodic in the sense that after each orbital revolution the vehicle returns to the same position with the same orbital speed. For the motion of a spacecraft around an irregularly shaped body, neither of these things is true. Figure 4.7 shows an orbit around an asteroid where the track of the spacecraft around the object does not join up, and in fact varies on each orbital rotation. This departure from ideal motion is particularly significant for close orbits around the object. Given that the orbit track is not repeated on each revolution around the object, it is possible that, after a number of orbits, the spacecraft may impact the surface. Care is required to choose a close orbit

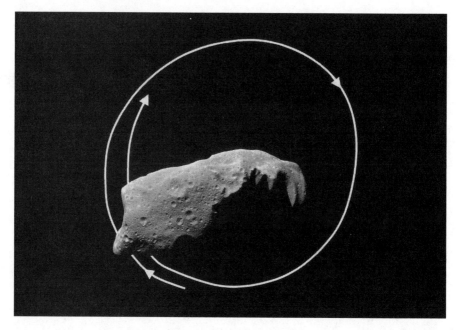

Figure 4.7: A close orbit around a small, irregularly shaped body is perturbed by the nonspherical distribution of mass in the asteroid. The picture is of the asteroid Ida, imaged by the Galileo spacecraft in August 1993. (Backdrop image courtesy of NASA/JPL-Caltech.)

that has long-term stability. On the other hand, if the spacecraft were to orbit some distance away, say 20 asteroid diameters, then the irregularities in the bodies shape become less influential, and the spacecraft's motion approximates well to an ideal orbit.

For missions like Rosetta, the process of deploying a lander on a comet can be quite torturous, given that the spacecraft has to be in fairly close orbit around the nucleus to successfully achieve this deployment. The first issue for the spacecraft designer is that the surrounding environment and the size of the comet are not known before the spacecraft gets there. This means that the spacecraft systems and payloads have to be designed to successfully and safely accommodate a range of conditions. The fact that the size, mass, and shape of the comet are unknown at the spacecraft design stage means that a gradual approach to a close orbit is required. As a consequence the spacecraft is inserted into an initial orbit with a large radius compared to the size of the comet.

For the sake of argument, let's suppose the comet nucleus is a fairly average 5 km (3 miles) in diameter, with a density about that of water. If we make the initial orbit radius equal to about 20 comet diameters, then the orbit would approximate well to an ideal circular orbit with a radius of around 100 km (62 miles). The speed of the spacecraft in this initial orbit is about 20 cm/sec (8 inches/sec), which compared to a brisk walking speed of around 4 mph or 1.8 m/sec (5.9 feet/sec) is an amazingly sedate speed for a spacecraft! This gives a first insight into the way mission orbit activities are carried out in proximity to a comet; it can be describe as a low-energy environment where things move slowly, and take a long time to happen. The other implication of this low orbital speed is the precision with which the entry into orbit has to be made. In this circular orbit the escape velocity is only 30 cm/sec (1 ft/sec), compared to about 11 km/sec for Earth, so an error in speed in excess of just 10 cm/sec in orbit insertion speed would result in the spacecraft's disappearing from the comet altogether!

The next job for the mission analysis team is to try to lower the orbit radius to, say, 2 comet diameters (10 km) to allow the lander to be detached from the spacecraft and begin its descent to the surface. However, to fly the spacecraft that close to an irregularly shaped comet would require the team to have detailed knowledge of the gravity field to facilitate accurate prediction of the spacecraft's trajectory. As mentioned above, without this knowledge it is possible over a number of orbital revolutions for the spacecraft to impact the surface. To acquire this knowledge, the comet is examined from the initial high orbit using imaging sensors to acquire detailed information about its shape. At the same time, the spacecraft is tracked precisely in order to detect the small perturbing forces that provide a clue to how the object's gravity field differs from an inverse square law. These data, on shape and perturbations, allow a first estimate of the gravity field of the comet to be made by the mission analysis team. With this information, the orbit radius can be lowered further to, say, 5 comet diameters (25 km), where the process can be repeated, allowing a further refinement in knowledge of the gravity field. Finally, once the team is confident that it has sufficient knowledge of the gravity field, the spacecraft can be inserted into its final close orbit, from which the lander can be deployed. The orbital speed in the 10 km radius orbit is around 60 cm/sec (2 feet/sec), still much slower than our brisk walking pace. In a real project situation, the choice of this orbit radius is difficult to pin down. On the one hand, it has to be small enough so that the descent time of the lander is not too long, but it also has to be large enough so that the orbiting spacecraft is not damaged or contaminated by the near-comet environment. The near-comet space is a dynamic environment, particularly when the comet is close to the Sun. Solar heating causes the surface of the comet to evaporate in plumes of gas and dust, from which the orbiting spacecraft needs to keep a distance. There is always a conflict between the engineers and scientists in these situations. The engineers want to keep the spacecraft a safe distance from hazards, whereas the scientists want it to be "where the action is." Clearly a compromise has to be struck between the value of the science and the risk to the spacecraft.

Let's suppose the lander is an instrument package with landing legs, having a total mass of, say, 100 kg. The next task is to detach it from the orbiting spacecraft, so that it can begin its descent to the surface. The simplest way of doing this is to push the lander out the back of the spacecraft, with an ejection speed equal to the spacecraft's forward orbital speed (how this is done is not really relevant to the story here, but it would probably be most easily done with a calibrated spring mechanism). If the backward speed of the lander matches the forward speed of the spacecraft—around 60 cm/sec in this case—then the lander is left momentarily motionless above the comet's surface. The comet's weak gravity will then cause it to fall steadily toward the surface. Because the gravity of the comet is so weak, this trip to the surface takes quite some time, around 4.4 hours in our example, and when the lander finally approaches the surface its descent speed is only around 1.6 m/sec (just a shade less than our brisk walking speed of 4 mph). On final approach to the surface, contamination of the surface by spraying it with rocket engine exhaust gases may not be a good idea from the point of view of the science objectives of the mission. This may prevent the use of a rocket engine to slow the descent, so that the structure of the lander may need to be able to withstand the 1.6 m/sec impact speed.

Perhaps the major issue with the touchdown is preventing the lander from bouncing back off the surface, and possibly going back into orbit. This is a real possibility, as the 100-kg-mass spacecraft will weigh only about $1/10^{th}$ of a Newton on the surface, which is about $1/10^{th}$ the weight of a small apple! And of course the nature of the surface, whether it's bouncy, or sticky, or somewhere in between, is unknown until touchdown. To prevent such a bounce happening, the lander will either have to fire an upward directed rocket motor or mechanically grapple the surface somehow, on touchdown.

Figure 4.8 shows the European Space Agency's Rosetta lander after what is hoped to be a successful touchdown on comet 67P/Churyumov-Gerasimenko in around 2014. The complexities of the orbital aspects of such a mission are not to be underestimated, where the central body is no longer a nice uniform sphere, but has a shape resembling a potato.

Figure 4.8: Artist's impression of the Rosetta lander after a successful touchdown on comet 67P/Churyumov-Gerasimenko. (Image courtesy of the European Space Agency [ESA].)

Halo Orbits Around Lagrangian Points

In earlier chapters, we described a variety of orbits around Earth, but in this section we discover that it is possible for a spacecraft to orbit around a point in space where there is no mass! This rather fascinating state of affairs requires some explanation.

The Three-Body Problem

The story starts in the 1770s with an Italian-French mathematician named Joseph Lagrange. Note that this is about 100 years after Newton gave the world his revelations about how objects move in gravity fields, giving the world's scientists and mathematicians potentially centuries of research work

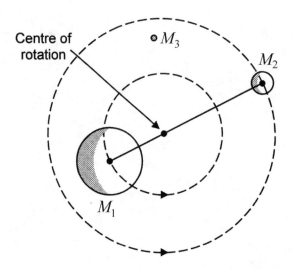

Figure 4.9: The circular restricted three-body problem. The two massive bodies M_1 and M_2 move in circular orbits around each other. The third body M_3 is of negligible mass and moves under the gravitational influence of M_1 and M_2.

to do to reveal their full significance. Lagrange was using Newton's laws to study something called the *three-body problem*, which, as the name suggests, is the investigation of how three massive bodies move around each other under gravity. Unfortunately, he found the problem to be complex and unsolvable, which it remains to this day. However, his work was not entirely fruitless. In his attempts to make the problem more amenable to solution, Lagrange examined simplified versions of the full three-body problem and in the process discovered *Lagrangian points*, which, as we will see, are relevant to modern spacecraft mission design.

The simplified version that Lagrange looked at is shown in Figure 4.9, and involves two massive bodies, M_1 and M_2, in circular orbits around each other, and a third much smaller body M_3 moving along a trajectory influenced by the gravity fields of its two larger neighbors. This setup is referred to as the *circular restricted three-body problem* (CRTBP). The important thing to note here is that the third body is so small in mass that it has negligible effect on the motion of the two larger bodies. So how do two massive bodies rotate around each other in circular orbits? If they are of equal mass, then they rotate about a point that is halfway between their centers (Fig. 4.10a). If their masses are dissimilar, they rotate about a point that is closer to the larger object (Fig. 4.10b). This point about which the rotation takes place is referred to as the *barycenter* of the system. For example, in the Earth–Moon system, the mass of Earth is about 81 times

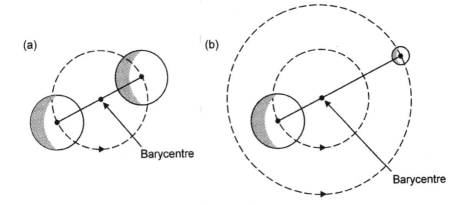

Figure 4.10: Massive bodies moving in circular orbits around each other rotate about their barycenter.

larger than that of the Moon, so the barycenter about which Earth and the Moon orbit is only about 5000 km (3100 miles) from Earth's center— actually beneath Earth's surface.

If we return to Lagrange's simplified problem, illustrated in Figure 4.9, there are a number of good examples of this type of system that are relevant to modern spacecraft mission design. The most obvious of these, from the 1960s, is an Apollo spacecraft on its way to the moon. In this example, Earth and the Moon represent the larger bodies in (nearly) circular orbits around each other, and the spacecraft represents the third body of negligible mass. Another example of a CRTBP with wide applications is the Sun–Earth– spacecraft system.

In his mathematical exploration of the CRTBP, Lagrange discovered five points in the rotating system where the third body (of negligible mass) could remain stationary relative to the two larger bodies. These *equilibrium points* are illustrated in Figure 4.11, and are referred to as Lagrangian points L_1, L_2, L_3, L_4 and L_5 in Lagrange's honor. When looking at Figure 4.11, it is important to realize that the relative geometry between the large bodies and the Lagrangian points remains fixed, and rotates about the system's barycenter.

Orbits About Lagrangian Points

What is the relevance of all this to spacecraft mission design? To focus the discussion, let's look at the situation where the two larger bodies are the Sun and Earth, and the smaller body is a spacecraft. Each Lagrangian point is a place where the forces of gravity and those due to the rotation of the system add up to nothing (we'll return in a moment to discuss what we mean by

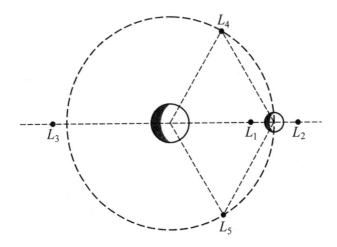

Figure 4.11: The locations of the Lagrangian points L_1, L_2, L_3, L_4 and L_5 relative to the two larger masses in a rotating system.

"forces due to rotation"), which gives it the characteristic of being an equilibrium point. If you locate a spacecraft at any one of these points, it will remain there. To further focus the discussion, let's consider the L_1 and L_2 points in the Sun–Earth system, as these have attracted most interest in terms of spacecraft applications. In this system, the L_1 point is located approximately 1,500,000 km (930,000 miles) away from Earth, in the direction of the Sun, and the L_2 point is a similar distance from Earth in the opposite direction (Fig. 4.11). Positioning a spacecraft at the L_1 and L_2 points seems like a straightforward affair, apart from one detail. A more detailed look at the mathematics tells us that these are points of *unstable equilibrium*, which means that if the spacecraft is disturbed by the slightest perturbation, it will move away from the Lagrangian points. This state of unstable equilibrium is a bit like trying to balance a small metal ball, like a ball bearing, on top of a smooth dome-shaped surface. With enough care, you may be able to balance the ball at the summit of the dome, but the slightest disturbance will cause it to roll away down the slope. And so it is with a spacecraft. However, with the help of the spacecraft's propulsion system, it is possible to regain stability, and furthermore to control the spacecraft in an orbit around the Lagrange point. How the spacecraft orbits a massless point is most easily explained by considering the motion about the L_1 point in the Sun–Earth–spacecraft system.

Strictly speaking, the L_1 point is a location in the rotating system where the gravitational and rotational forces sum to zero. I think we are fairly happy thinking about the forces of gravity of the Sun and Earth acting upon the spacecraft, but what do we mean when we talk about rotational forces? The Sun–Earth system rotates only slowly, about 1 degree per day due to the Earth orbiting the Sun with a period of 1 year, so whatever forces there are due to rotation, they must be small. But in this instance, they do nevertheless play an important role.

Perhaps the best way to think about rotational forces is to imagine yourself on a small merry-go-round, the kind you see in a child's playground. Maybe as a child you stood on one of these, holding on to the safety rails, while a friend spun it up to perhaps an uncomfortably high speed. In this situation, you certainly get a good impression of a rotational force, as this is the force you feel tending to throw you off the merry-go-round. This outward directed force you experience in a rotating system—the merry-go-round in this case—is referred to as *centrifugal force*. To prevent yourself from being hurled off the merry-go-round, you have to hold on tightly to the safety railings. You are able to remain standing on the same spot on the merry-go-round because the force in your arms pulling toward the center of the merry-go-round balances the centrifugal force tending to throw you off.

As it happens, this is a remarkably good analogy to describe the manner in which the spacecraft can remain "standing on the same spot" at the L_1 point in the rotating Sun–Earth system. The force tending to pull the spacecraft toward the Sun is the Sun's gravity, and this is analogous to the force in your arms as you hold on tightly to the merry-go-round's safety rails. The force tending to pull the spacecraft outward is predominantly centrifugal, generated by the rotation of the system, but there is also a small contribution from Earth's gravity. There is a balance of forces on the spacecraft—solar gravity inward, and centrifugal force plus Earth gravity outward—allowing the spacecraft to remain stationary at the L_1 point. To be precise, 97% of the outward force balancing solar gravity is centrifugal, and only 3% is Earth's gravity.

But we still have not addressed how a spacecraft can orbit the L_1 point. In Figure 4.12 we see that if the spacecraft is displaced along the line joining the Sun and Earth to point 1 or point 2, then the sum of gravity and rotational forces is no longer zero, and the vehicle will tend to move away from the Lagrangian point. This is an expression of the instability of the L_1 point that we referred to earlier.

However, there is a *surface of stability* upon which the spacecraft can orbit the L_1 point. This is the surface at right angles to the Sun–Earth line, which passes through the L_1 point, as shown in Figure 4.12. It is slightly curved, as shown in a rather exaggerated way in the figure, but it can be thought of as a plane surface in which the orbital motion takes place. Now we can see that if

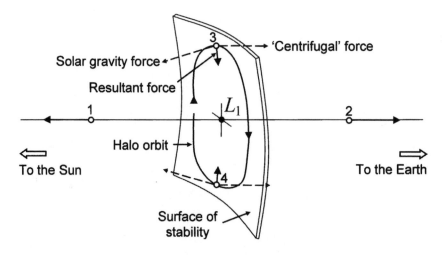

Figure 4.12: An illustration of how a spacecraft can orbit a massless Langrangian point.

the spacecraft is located at point 3 or point 4, the sum of the Sun's gravity force in one direction, and centrifugal force (with a little bit of Earth gravity) in the other, produces a resultant force directed toward the L_1 point. In fact, it is easy to see that this L_1 directed force occurs at any point on the illustrated halo orbit in the figure, thus allowing an orbital motion around the massless L_1 point. The shape of the orbit around L_1 is certainly not a conic section, in general, and can in fact be a weird variety of looping curves referred to as a *Lissajous orbit*. Also, as we have mentioned already, the spacecraft has to use its propulsion system to tweak the motion to ensure long-term stability of the orbit.

A similar explanation of the orbital motion around the L_2 point can be argued, with a balance between an inward directed force composed of Sun and Earth gravity and a outward directed centrifugal force. The latter force is slightly larger at the L_2 point since it is further away from the center of rotation.

As for the idea of a *resultant force*, the configuration of force vectors at point 3 in the figure is similar to the forces acting on an arrow when it is fired from a bow. In Figure 4.13 the force vectors actually acting on the arrow are the tension forces in the string on either side of the arrow. But the sum of these—the resultant force—is actually directed along the arrow, and produces the acceleration that makes it fly.

Why are we interested in this concept? The idea of using Lagrangian point orbits for spacecraft is not a new one. A halo orbit around the L_1 point, between Earth and the Sun, is an ideal location for a spacecraft with a Sun-

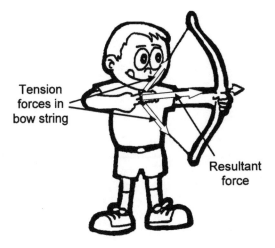

Tension
forces in
bow string

Resultant
force

Figure 4.13: The tension forces in a bow string produces a resultant force along the arrow.

viewing payload, as such a spacecraft has an uninterrupted view of the Sun. Examples of solar observatory spacecraft that have resided at L_1 are the SOlar and Heliospheric Observatory (SOHO) and the Advanced Composition Explorer (ACE). Conversely, the L_2 point is further away from the Sun than the Earth, above Earth's night side, and this point is a good location for space telescopes. The sky is not obscured by Earth as it is for the Hubble Space Telescope in its low Earth orbit; Earth subtends an angle of only about half a degree from the L_2 point. The Wilkinson Microwave Anisotropy Probe (WMAP) spacecraft is an example of a space observatory that has used a L_2 point orbit. Looking to the next generation of space telescope beyond Hubble, the James Webb Space Telescope (JWST) is destined for a L_2 halo orbit around the year 2013.

Bored with Orbits?

We seemed to have spent quite a time talking about orbital motion, one way or another, over the last four chapters. In Chapter 5 let's take a break and have a look at how we use rocket-powered launchers to get our spacecraft off the ground and into orbit.

Getting to Orbit 5

Rocket Science and Engineering

THE way rockets work is still a bit of a mystery to most people. A common misconception is the belief that they do not work unless the rocket exhaust has something to push against. I still have a vivid memory of this myself as a child; it was obvious to me that the early rockets taking the first pioneering astronauts into orbit rose majestically from their launching pads only because they were able to push against the ground and atmosphere. Clearly, my powers of thought did not stretch to face the dilemma of what happened once the rocket was whizzing about in space, with nothing to push against!

Another false impression in my view is the idea that rocket science is something that is really complicated. As we know, the phrase itself has become a byword in colloquial English for something that is complex and difficult to understand. For example, a neighbor of mine finds herself living between me on the one side and a hospital surgeon on the other, and recently commented that she felt wedged between a rocket scientist and brain surgeon! However, it is my humble opinion that the business of brain surgery is a lot more complicated, and in this chapter I hope to convince readers that the science of rockets is conceptually simple.

On the other hand, the engineering of rockets is another matter entirely. The rocket engine poses major challenges to the engineers who wish to transform the simple concept into high performance hardware. We can get a feeling for this challenge if we think about what happens each time a space shuttle is launched into Earth orbit. The shuttle orbiter's mass is on the order of 100 metric tonnes, and needs to be accelerated to a speed of approximately 8 km/sec (5 miles/sec) to reach orbit. When you consider that the amount of energy needed to do this is equivalent to around 700 metric tonnes of the high explosive TNT, you begin to appreciate the magnitude of the problem. This amount of energy has to be released gradually and in a controlled manner over a period of minutes by the rocket engines, and the consequences of a mistake are catastrophic. The processes involved in this controlled release of energy places the rocket hardware

G. Swinerd, *How Spacecraft Fly: Spaceflight Without Formulae,*
DOI: 10.1007/978-0-387-76572-3_5, © Praxis Publishing, Ltd. 2008

under significant mechanical and thermal stresses that are difficult for the engineers to manage.

This is why the business of launching spacecraft into orbit is such a risky business. For most current launcher systems, a *reliability* of 90%, that is, nine flights in 10 are successful, is considered to be acceptable. With the most reliable launch system, the space shuttle, the reliability increases to around 99%, which is considered to be very good for a conventional launch system. However, if you think of it in terms of the reliability of civil air transport, clearly space transportation has a great deal of room for improvement. If you were told that you had one chance in 100 of not reaching your holiday destination on an airliner, my guess is that you would probably stay at home!

Many people have become rather blasé about the whole business of manned space flight, but when the chips are down, the men and women who do this are literally staking their lives on the successful operation of a highly stressed piece of rocket hardware. If you are a young scientist or engineer, and wish to make your name and fortune, then a new and safer way of achieving orbit is a problem worth addressing. But it is a rather difficult one to crack; perhaps one of those "Beam me up, Scotty" machines that some of us have grown up with on *Star Trek* would be a contender!

Replacing Newton's Cannon

In Chapter 2, we discussed the nature of orbital motion using Newton's cannon as a means of achieving orbit. Clearly we do not have a handy 200-km-high (124-mile-high) mountain upon which to build such a structure, so instead we resort to rocket-powered launch vehicles to place our satellites into orbit. It is fairly easy to see how the substitution is made. As we saw earlier, to enter a low orbit around Earth, a cannonball needs to be traveling horizontally out of the cannon's barrel at around 8 km/sec (5 miles/sec) at an altitude of 200 km. Thereafter, we saw that the curvature of the cannonball's trajectory matched that of Earth's surface, thus avoiding ground impact and ensuring the cannonball's orbital state. If we wish to launch a satellite installed on top of a launch vehicle into the same orbit, then we have to ensure that the same *initial conditions* are met; if the launcher can lift the satellite to an altitude of 200 km and boost its speed to the required 8 km/sec in a horizontal direction, then the same orbital motion will result.

The launch vehicle is more versatile than the cannon, as the latter is somewhat constrained in its performance by being fixed to its supporting platform—the mountain—at a 200-km altitude. If the launcher is powerful enough, it can lift the satellite to higher altitudes, and vary the satellite's initial speed and direction to enter a variety of orbits as discussed in Chapter 2. For example, if the launch vehicle were to release its payload at a height of 700 km (435 miles) in a horizontal direction with a speed of 7.5 km/sec (4.7 miles/sec), the satellite will enter a circular orbit at that altitude. Furthermore, if the satellite was to be released in the same way at this height with a speed of 10.6 km/sec (6.6 miles/sec), then it would eventually escape Earth on a parabolic trajectory (see Chapters 1 and 2).

It *Is* Rocket Science …

The surprising thing about a rocket engine, as we have already intimated, is that the concept is not at all complicated. It does not have complex mechanisms like pistons going up and down as in an automobile engine. Instead, it comprises only three basic parts: a *propellant feed system*, a *combustion chamber* where the fuel is burned, and a *nozzle* out of which the resultant hot gases are exhausted. And that's it.

If we're thinking about rocket engines used on launch vehicles, then it is obvious that they need to be big. For example, all the components of a space shuttle sitting on the launch pad have a combined mass of about 2000 metric tonnes! The rocket engines—five of them in this case—have to produce enough thrust between them to match the vehicle weight, and then a bit more to move it vertically off the pad, as beautifully illustrated in Figure 5.1. Two types of rocket engine are used by the shuttle to gain orbit, the *solid propellant motor* and the *liquid propellant motor*, and these are the most commonly used systems on launchers.

The *solid propellant motor* (Fig. 5.2a) is a bit like a giant firework, inasmuch as you light it, and it burns to produce thrust until the solid propellant is exhausted. To burn the fuel in any rocket system, we need oxygen. In an airplane, the fuel is burned using oxygen taken from the atmosphere through the engine intakes. However, since there is no atmospheric oxygen available as the launcher approaches orbital altitudes, the rocket system has to take its own oxygen with it. For a solid propellant motor, the fuel and oxidizer are mixed together in a gooey substance, which is then poured into molds to set hard. Often the molds are shaped to produce sections of solid propellant like giant disks with a hole punched through the middle, which are then stacked in the rocket's cylindrical casing to produce the configuration shown in Figure 5.2b. The rocket is then lit by a *pyrotechnic device*, usually at the top of the rocket, and the propellant burns from the inside of the cylinder outward toward the metal casing, until depleted.

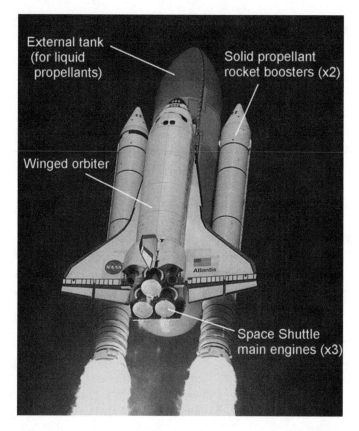

External tank
(for liquid
propellants)

Solid propellant
rocket boosters (x2)

Winged orbiter

Atlantis

Space Shuttle
main engines (x3)

Figure 5.1: The components that make up the space shuttle launch system. (Image courtesy of the National Aeronautics and Space Administration [NASA].)

The space shuttle uses two large solid propellant rocket boosters for its first two minutes of flight, which use a combined fuel and oxidizer solid propellant. When the system was designed in the 1970s, a decision was made to use solid propellant boosters, mainly to constrain costs. Many of the rocket scientists involved were unhappy about their use on a system designed to carry people. The basis of this concern was the fact that solid rockets are less controllable than their liquid propellant counterparts, the main worry being that once a solid rocket is ignited, it cannot be stopped until the fuel is exhausted. The boosters used on the shuttle have a large thrust, and have to be ignited simultaneously at the moment of lift off. If one were to fail to ignite at that critical moment, the resulting asymmetry in the vehicle's thrust would be catastrophic.

The *liquid propellant rocket motor* (Fig. 5.3) is a little more complicated, requiring a propellant feed system, in addition to the combustion chamber

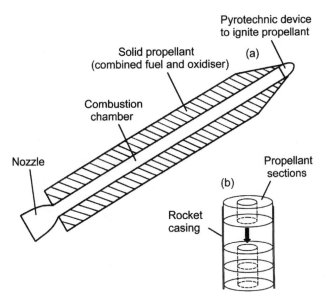

Figure 5.2: (a) The elements comprising a typical solid propellant rocket. (b) The assembly of the solid propellant sections in the rocket casing.

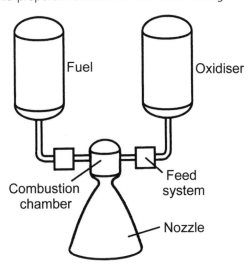

Figure 5.3: Schematic of the components of a liquid propellant rocket motor.

and rocket nozzle. In this case, the launcher carries its fuel and oxidizer in separate tanks, with a feed system—usually pumps—to shift the liquids into the combustion chamber, where they are ignited to produce a highly pressurized hot gas that exits through the nozzle of the engine to produce thrust. As we will see in a moment, the faster the exhaust gases exit the nozzle, the better. The internal cross section of the nozzle has a particular profile, first converging to form a *throat* and then diverging to form the familiar bell shape. The exhaust gases accelerate as they squeeze through the throat, typically reaching the local speed of sound. Thereafter, the gases continue to accelerate and expand in the *divergent section*, ensuring a high *exit speed* and ideally an *exit pressure* near to the ambient atmospheric pressure.

Since the propellant can be supplied to the combustion chamber in a controlled manner through the feed system, liquid propellant rockets are inherently more controllable, allowing the thrust level to be varied and the rocket to be turned on and off if required. There are a variety of fuel and oxidizer combinations, but commonly used ones are *hydrogen/oxygen* and *kerosene/oxygen*.

In addition to the two solid propellant boosters mentioned above, the space shuttle also requires the use of three liquid propellant rocket motors to acquire orbit; these are referred to as the *space shuttle main engines* (SSMEs). It is these liquid-powered rocket engines that continue to operate to take the vehicle to orbit, once the solid propellant boosters are depleted and fall away. The SSMEs use a combination of hydrogen and oxygen as fuel and oxidizer, respectively. These gases need to be cooled to very low temperatures to produce a *cryogenic liquid*, and these are stored in the large insulated external fuel tank prior to launch (see Fig. 5.1).

Action and Reaction

Why does exhausting a hot gas through a nozzle produce a force that propels the rocket vehicle to high speeds? Once again, Isaac Newton comes to the rescue in answering this question.

There are two main thrust effects occurring to allow the rocket vehicle to accelerate in the atmosphere, and in the vacuum of space where there is nothing to push against! They are referred to as pressure thrust and impulse thrust, the latter being by far the dominant source of thrust force in a rocket system.

The way that *pressure thrust* works is easily understood by thinking about a bottle of carbonated soda (Fig. 5.4). When the bottle is sealed, the pressure forces inside are equal on all sides of the container, and there is no resultant force. However, if we remove the cap, an imbalance occurs producing a net force (which is small in this case). The bottle with the cap removed is analogous to the rocket engine combustion chamber with the open exit nozzle; there is some propulsive force produced by the pressure imbalance.

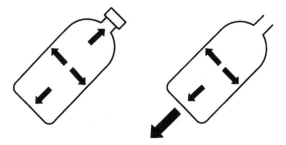

Figure 5.4: A pressure imbalance produces pressure thrust.

To understand the dominant propulsive effect, the *impulse thrust*, we need to recall Newton's laws of motion discussed in Chapter 1, and in particular his third law: to every action there is an equal and opposite reaction. Put simply, the action of a rocket engine is to throw lots of mass at high speed out the back of a launch vehicle, and the reaction is to cause the vehicle as a whole to accelerate in the opposite direction. To see this more clearly, we can propose an experiment to demonstrate the principle, but, again, do not try this at home! All you need for this is a high-velocity rifle, a skateboard, and a good sense of balance to stop you falling off the skate board when you fire the gun! A fired rifle produces a "kick": when the trigger is pulled, the bullet flies out of the barrel at high speed in one direction, and the gun reacts alarmingly by kicking back on the shooter's shoulder. It is this kick that can be harnessed as a propulsive force if we fire the gun while standing on the skateboard (Fig. 5.5). Let's suppose that the combined mass of the shooter, the rifle, and the skateboard is 75 kg. Let's suppose that the bullet has a mass of 50 g and leaves the barrel of the gun at 1500 m/sec (4920 feet/sec). We can

Figure 5.5: A "thought experiment" involving a high-velocity rifle and a skateboard!

do some simple calculations to show that if the shooter gets on the skateboard and fires the gun, the kick will give them a speed of 1 m/sec (3 feet/sec) in the other direction. That's just over 2 mph, which is less than a good walking pace, so our improvised rocket system is not that good at producing propulsive force. But it does illustrate the effect of impulsive thrust. In a real rocket system, the designers strive to eject lots of "bullets"— in this case the *exhaust gases* from the engine nozzle—at high speed to maximize the reaction to accelerate the vehicle in the opposite direction.

Note that the two sources of rocket thrust are closely related. Clearly, the hole in the combustion chamber where the nozzle exit is located causes a pressure imbalance, giving a measure of pressure thrust. Also, equally clearly, if you have a hole with high pressure gas inside and low pressure gas outside, then you are definitely going to get mass coming out, providing impulse thrust. Although the two effects are distinguishable in the mathematics of rockets systems, physically they are closely linked.

Some Fundamental Propulsion System Numbers

There are principally two numbers that rocket scientists use to assess a rocket's performance and to compare one system with another: thrust and specific impulse.

Thrust is a fairly intuitive idea, and is basically the level of force provided by the rocket to accelerate the vehicle. You may recall from Chapter 3 (see the discussion on aerodynamic forces on spacecraft) that the thrust force is usually measured in Newtons. A Newton of force has a formal definition, but you may remember our approximate and informal definition, which is that it is about the weight of a (smallish) apple! Another helpful measure, when thinking about the large forces associated with launcher engines, is that a metric tonne weight is about 10,000 Newtons. Again, in the space shuttle system each of the three SSMEs has a thrust of the order of 2 MN (a MN is a MegaNewton, which is 1,000,000 N), and the thrust of each of the two solid propellant boosters is approximately 10 MN.

Specific impulse, measured in units of seconds, is not an intuitive concept, but it is an important one since it gives a measure of how much change in speed can be achieved by a rocket system for a given mass of propellant. For example, a liquid propellant rocket motor using hydrogen/oxygen as a fuel/oxidizer combination has a typical specific impulse of 450 seconds, whereas a solid propellant rocket has a value around 250 seconds. To see the relevance of the specific impulse, we can envisage two rocket vehicles with the same initial mass, one powered by the liquid propellant motor and one with a solid motor. If each rocket burns exactly the same amount of propellant, the change in speed achieved by the liquid propelled vehicle will be 450/250, that is ~1.8 times the speed change achieved by the solid propelled vehicle. The reason why the liquid powered rocket does so much better is that it is ejecting exhaust gases out of its rocket nozzle at a much higher speed than that achieved by the solid propellant motor. This observation provides another, perhaps more physical interpretation of the specific impulse parameter—a high specific impulse implies a high exhaust velocity. Generally speaking, the higher the specific impulse, the better.

Some More "Gee-Whiz" Information About the Space Shuttle

The rocket systems that we have discussed are fundamentally simple in concept, but as we have already said at the beginning of this chapter, it is the engineering implementation of these simple concepts that represent the major challenges for launch vehicle designers and manufacturers. The SSMEs, with a specific impulse of around 450 seconds, are considered to be the best you can do currently for a chemical propulsion system. They are designed to be reusable, which is extraordinary when you consider the mechanical and thermal stresses involved in their 8 minutes of operation from ground to orbit.

The shuttle has three liquid-fueled SSMEs, and the description of the attributes of just one of these fills me with admiration for the engineers who have transformed the concept into reality! The combustion chamber operates routinely at a temperature of around 3300°C, which is approximately twice the melting point of steel. To get the high exhaust velocity, the combustion chamber pressure is equivalent to about 200 times atmospheric pressure. With this magnitude of pressure, the propellant feed system has to be substantial to be able to push the fuel and oxidizer into the chamber against that pressure. The turbopumps that perform this function rotate at about 37,000 rpm to provide chamber inlet pressures of 305 atmospheres for the liquid oxygen and 420 atmospheres for the liquid hydrogen, with a total fuel flow rate of 470 kg/sec—nearly half a metric tonne a second! The resulting thrust level is around 2 MN, with a nozzle exhaust velocity of roughly 4500 m/sec (14,800 feet/sec).

At the moment of takeoff, the space shuttles engines are throwing an awful lot of hot gas down into the flame trench of the launch pad, about one and a half metric tonnes per second of exhaust gases, traveling at about 4500 m/sec, from the three SSMEs, and about 8 metric tonnes per second at a speed of 2500 m/sec (8200 feet/sec) from the two solid propellant boosters— a little bit more of a kick than our 50-g high-velocity bullet traveling at 1500 m/sec! There is clearly a lot of destructive power here that needs to be managed to prevent damaging the launch pad. To deal with this, what might be called the "space shuttle swimming pool" comes to the rescue. In a recent IMAX big screen 3D film of the International Space Station, one spectacular sequence showed the launch of a space shuttle at close quarters. The controlled power of the vehicle was overwhelming to watch! But one striking thing that came over clearly was the cascade of water that is released into the flame trench from the "swimming pool" just prior to ignition of the SSMEs. The energy in the high-speed jets of hot gas from the engines is consumed in the process of converting all that water into steam rather than causing significant damage to the launch pad.

Ascent to Orbit

There are a variety of types of launch vehicle, ranging from expendable to totally reusable. However, most of the traffic to orbit currently is in the form of *expendable launch vehicles* (ELVs). As the name implies, these vehicles are used only once, and their various components are jettisoned on the ascent or abandoned in orbit. In *semi-reusable launch vehicles*, of which the space shuttle is the only current example, some parts of the vehicle are destined for reuse on subsequent flights, such as the winged orbiter and the solid propellant booster casings (see Fig. 5.1). Other parts such as the huge orange external fuel tank are jettisoned on the ascent, and subsequently reenter the atmosphere and burn up over, or plunge into, the ocean.

In *reusable launch vehicles*, of which there is are no current examples, the launcher is operated somewhat like an airplane to significantly reduce operational costs. The development of such a launch system is the "Holy Grail" of rocket scientists at the moment, but, as we will see later, the technical challenge is significant, and the likely development cost a deterrent! It seems probable that this will first be achieved under the umbrella of a military program, where financial resources are perhaps less of an issue. The description *man-rated* can also be applied to any of the three types mentioned above, which implies that the launch vehicle has been developed to carry people into orbit; for example, the shuttle can be referred to as a man-rated semi-reusable launch vehicle. Applying the man-rated label to a launch vehicle has major technical implications, as the reliability must be significantly better than the 90% or so, typical of an unmanned ELV system. And the required improvements in reliability will cause a significant increase in the launch vehicle's development costs.

Dynamics of the Launcher

We now turn to the dynamics of the launcher as it rises to orbit. The launch of a spacecraft is conducted in a particular way, in order to ensure that the launch vehicle can carry a good payload mass into orbit. Three of the most commonly used strategies to attempt to maximize the launcher's payload mass are discussed in the following subsections.

Staging

A typical expendable launch vehicle adopts the method of *staging* to reach orbit, which involves shedding mass on the ascent. The vehicle is made up of stages, usually three, as illustrated in Figure 5.6. The launcher is lifted from its pad using the first-stage engines. The vehicle then climbs and accelerates until the first-stage fuel is exhausted. At this point, the first stage separates and falls back to Earth; there is no need to lift the first stage to orbit since it would serve no purpose there. It is jettisoned to save the expenditure of precious propellant in accelerating useless mass to orbit. After separation, small rocket thrusters are often fired in the first stage to slow the rocket down and move it away from the second-stage engines. At this point, prior to the ignition of the second-stage engines, the upper stages and satellite payload are in a state of free-fall—like the hypothetical elevator discussed in Chapter 2. As a consequence, the second-stage propellant and oxidizer are effectively in a state of weightlessness in the tanks. To settle the liquids again at the base of the tanks, in order to feed them into the combustion chamber, small rocket thrusters attached to the second stage are fired to speed it up. This acceleration produces a bit of artificial gravity to aid this necessary management of the propellant. The second-stage main engines are then fired to continue the powered ascent to orbit. When the second-stage fuel is used up, the process is repeated; the second stage too is jettisoned, before the third-stage engine is lit to take the satellite payload to orbital speed and altitude. It is often the case that the third stage and satellite payload are both injected into the final orbit.

Staging is a vital strategy in the operation of a conventional expendable launch vehicle. Fuel mass is not wasted accelerating rocket components to orbit where they would be useless, but instead is used to maximize the mass of the spacecraft being launched. Also the change in speed required to reach orbit is acquired using conservative engine technology in each stage. Rocket

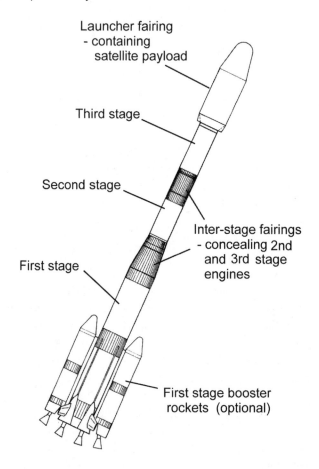

Figure 5.6: The components of a typical three-stage expendable launch vehicle. (Backdrop image courtesy of Arianespace.)

engines having lower specific impulses can be used (lower certainly than the high performance space shuttle main engines), which implies lower combustion temperatures and pressures. This in turn means that the level of mechanical and thermal stress imposed on the engine components is generally less, which is good news for engine reliability and costs.

Ascent Trajectory Optimization

Another aspect ensuring that the launch vehicle can carry a good payload mass into orbit is the optimization of the ascent trajectory. This is a mathematics- and computer-intensive activity, usually carried out by the launch agency, to minimize the amount of rocket propellant needed to reach the desired orbit. If we achieve this minimization of fuel mass, then it follows that we can invest the resulting mass savings in making the spacecraft payload that the launcher it is carrying bigger and more capable. Clearly, the ascent trajectory optimization is an important task, and if not done correctly it can seriously compromise the vehicle's launch capability.

Despite the mathematical complexities of the process, we can gain some understanding of the ascent optimization by thinking about the tasks achieved by the launcher's propellant on the ascent. First, and most obviously, propellant mass is used to gain speed; the launcher starts out with effectively zero speed on the launch pad, and has to be accelerated to around 8 km/sec (5 miles/sec) to achieve a low Earth orbit (LEO). Second, and perhaps less obviously, propellant mass is used to overcome the forces of gravity and aerodynamic drag, which act on the launcher during its ascent.

Expending propellant to overcome gravity is referred to as *gravity loss.* This idea is nicely illustrated by a vertical takeoff aircraft, like a Harrier jump jet, when it hovers just above the ground. In this state, it is using its fuel entirely for the purpose of overcoming the force of gravity. Similarly, if a launcher's ascent trajectory has an uphill slope—and yes, of course, it would have to in order to reach orbit—then some part of its fuel is being used the overcome the force of gravity. When the launch vehicle is initially climbing vertically from its launch pad, the gravity loss incurred is large, but if it can roll over into a more gently sloping flight path soon afterward, the gravity loss is reduced.

Propellant is also used to overcome aerodynamic forces (see also the discussion of aerodynamic forces acting on spacecraft in Chapter 3), principally drag, acting on the launch vehicle during the ascent. This is referred to as *drag loss.* We can feel the effect of aerodynamic drag by holding a hand out of an open car window on a summer's day; the flow of air produces a force on your hand that resists its motion through the air. By this simple means we can also get an idea of how the drag force varies with speed. For example, the drag force on your hand feels much more significant at 60 mph than at 30 mph. If you were able to measure it, you would find the force at 60 mph to be four times bigger than the force at 30 mph, suggesting that the drag force varies as the square of the speed. As the launcher's speed doubles, the drag force increases by a factor of 4 (2^2), and as its speed trebles the drag increases by a factor of 9 (3^2), and so on. Given the speed that the launcher ultimately attains to reach orbit, this sounds like bad news, but the saving grace is that the drag loss occurs only in the lower, denser part of the atmosphere from which the launcher can escape fairly quickly.

The magnitudes of the gravity and drag losses are dependent on the launch vehicle being used, but typically if we are having to accelerate to a speed of 8 km/sec to reach orbit, we would have to burn propellant equivalent to an additional 1.0 to 1.5 km/sec to overcome gravity, and something like an extra 0.3 km/sec to combat the effects of drag.

Returning to our discussion of optimization of the ascent trajectory, we can now see that we want to design the flight path to ensure that most fuel is used to acquire speed, and that the amount used to overcome gravity and drag losses is minimized. One way of acquiring orbit is to ascend vertically to orbital height, and then rotate the launcher's flight path into the horizontal direction to inject into orbit. Although this strategy will minimize drag loss—the vehicle climbs vertically through the denser part of the atmosphere quickly—it does, however, entail accumulating a huge gravity loss. Alternatively, the launcher can climb into orbit on a gently sloping trajectory, like that of an airplane, with a small climb angle. In this case the gravity losses would be minimized, but the vehicle would spend a long time climbing out of the denser part of the atmosphere, thus yielding a large drag loss. An optimized trajectory, therefore, tries to take a path between these two extremes. Consequently, the launcher will climb vertically for a relatively short period to escape the denser part of the atmosphere (to minimize drag loss), and then roll over into a shallow climb (to minimize gravity loss) to acquire orbit. Figure 5.7 shows a space shuttle adopting this strategy.

Using Earth Rotation

As well as staging and trajectory optimization, a third way of improving the mass that a launch vehicle can take to orbit is to use Earth rotation. To get a feeling for this, we note that when the launcher is just sitting on its pad doing nothing, it is already moving eastward at significant speed due to the fact that Earth is rotating. The only places where this is not true are the North and South Poles, and I am not aware of any launch facilities in these polar regions, apart from perhaps submarines in the Arctic Ocean! The magnitude of this speed depends on the location of the launch site, or more precisely on its latitude. If the site is located on the equator, at zero latitude, then by virtue of Earth rotation it is traveling at about 465 m/sec (1525 feet/sec) in an eastward direction. This speed decreases as we move away from the equator; for example, at a latitude of 28 degrees north, corresponding to the Cape Canaveral launch site in Florida, the launch pads are moving at a rate of around 410 m/sec (1350 feet/sec). Further north at a latitude of 60 degrees, the eastward movement of Earth's surface has reduced to half of that at the equator, and at the North Pole it has reduced to zero. As I write this in southern England at a latitude of 52 degrees, it is amazing to think that I am rushing eastward due to Earth rotation at about 285 m/sec (940 feet/sec), but I cannot feel a thing!

Figure 5.7: A space shuttle launch, illustrating an optimized ascent strategy. (Image courtesy of NASA.)

Getting back to our launch vehicle, we can see that if we take off vertically but then roll over and head down range in an easterly direction, we can take advantage of the effect of Earth's rotation. For example, a rocket launched from my garden is already moving east at 285 m/sec before I light the blue touch paper, so if it is guided down range toward the east, it will burn less fuel to reach orbital speed. This means that the saving in propellant mass can be used to increase the size of the satellite payload being lifted to orbit. Another consequence of this eastward-directed launch strategy is that the orbital inclination (see Chapter 2) of the resulting orbit is about the same as the latitude of the launch site. The geometry of this is illustrated in Figure 5.8. Consequently, a space shuttle launched in this way from Cape Canaveral at a latitude of 28 degrees will end up in a LEO inclined at 28 degrees to the equator.

A result of all this is that large spacecraft, for example, the Hubble Space Telescope, the Space Shuttle, and the International Space Station, orbit in low-inclination LEOs. The savings in fuel usage as a result of using Earth rotation allow generally larger payloads to be launched into this type of orbit. Also launch sites for satellites destined for geostationary Earth orbit (GEO) (see Chapter 2) are usually sited near the equator with good range safety to the east. The safety requirement usually means no human habitation beneath the flight path, which is often satisfied by having an expanse of ocean to the east of the launch site. A good example of this is the launch facility at Kourou in French Guyana, at a latitude of 5 degrees north,

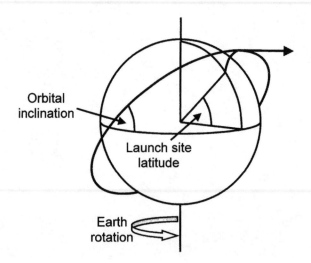

Figure 5.8: A launch in an easterly direction results in an orbit with an inclination approximately equal to the latitude of the launch site.

from where the Ariane family of expendable launch vehicles are launched predominately, eastward over the Atlantic Ocean.

But rockets are not always launched toward the east. For example, if the inclination of the mission orbit is required to be near 90 degrees, then the launch vehicle would have to fly either north or south from the launch site to achieve this. Therefore, Earth rotation is of no benefit, and reaching orbit is more costly in terms of fuel mass. Consequently, this increase in fuel mass must be compensated for by having a lower satellite payload mass, all other things being equal. Generally speaking, launcher performance is reduced in terms of payload mass when, for example, a near-polar mission orbit is required.

Launch Vehicle Environment and Its Effects on Spacecraft Design

Another attribute of the launch process, which has a major effect on the spacecraft design, is the *launch vehicle environment*. Usually the structural design of the spacecraft (see Chapter 9) is not governed by its lifetime in orbit, which could be 10 years, but rather by the few minutes it spends climbing to orbit on the launch vehicle. We have already seen how much energy has to be released in a controlled manner during this relatively short period of time to achieve an orbital state for the spacecraft. It is not too surprising, therefore, to realize that the spacecraft is exposed to a lot of *noise* and *vibration*, and is subject to significant levels of *acceleration* to boost its speed from effectively zero to 8 km/sec (5 miles/sec) in a few minutes to reach orbit.

For launch spectators, one overriding impression is the wall of sound that hits them a few seconds after they see the rocket engines ignite, despite the fact that they are kept at a safe distance from the launch complex. Launch is a very noisy affair, and even more so for the satellite payload sitting on top of the rocket. The acoustic field encountered by the satellite is harsh, despite the satellite being contained within the launcher fairing. Large amplitude and damaging vibrations can be excited in flexible structures, such as solar panels or large antennas, by this level of noise.

Similarly, it is not surprising that the energetic processes occurring in the propellant feed pumps, the combustion chambers, and the rocket nozzles at the base of the launch vehicle cause a high level of vibration. Astronauts riding a man-rated launch vehicle to orbit invariably report quite a rough ride!

In addition to noise and vibration, the third main environmental effect to which the spacecraft is exposed during launch is acceleration. Figure 5.9 shows a typical acceleration profile against time for an Ariane 5 launch vehicle, where the acceleration is given in units of *g*'s. To understand the impact this has on the spacecraft design, we need to recall the discussion in Chapter 1 about the force of gravity at Earth's surface. If you drop something, it will accelerate toward Earth's center, increasing its speed by 10 m/sec (32 feet/sec) for every second it falls. This acceleration of 10 m/sec/sec, usually denoted by 10 m/sec^2, is the reason we stay stuck to the floor. This environment, in which we experience our normal weight, is sometimes called a 1*g* environment. However, when we ride to the top of a skyscraper in a high-speed elevator, while the elevator is accelerating upward to gain speed we feel heavier (see Fig. 2.2b). And so it is with launch vehicles. Figure 5.9 shows that the level of acceleration for this particular launch vehicle can be in excess of 4*g*—four times Earth's surface gravitational acceleration— which means that the spacecraft and its component parts effectively weigh four times their normal Earth surface weight. It is not too difficult to see the effect this has on the spacecraft structure, since its job is to mechanically support all the various parts of the spacecraft—payload instruments and subsystem elements—which are effectively much heavier under severe acceleration.

Clearly, the spacecraft structure design engineer has to take into account the launch vehicle flight environment to ensure that the spacecraft does not fall apart on the ascent to orbit.

Next-Generation Launchers

Perhaps the most striking thing about launching spacecraft using conventional ELVs is how inefficient and costly it is. Typically only about 1% of the mass that sits initially on the launch pad reaches orbit and is usefully employed to fulfill the mission objective. The remaining 99% is jettisoned either on the ascent or in orbit. What can be done about that? Well, a great deal. Rocket scientists want to develop a launch vehicle with operating characteristics similar to a civil aircraft: a launcher that takes off from a conventional runway, delivers a payload to orbit, and returns to a runway landing without jettisoning big lumps of itself on the way. If this can be achieved, then the cost of access to orbit would be significantly reduced, which would accelerate the exploitation of space in the existing areas of application satellites and scientific research. If this revolution can be achieved in a way that increases the reliability of launchers to match civil

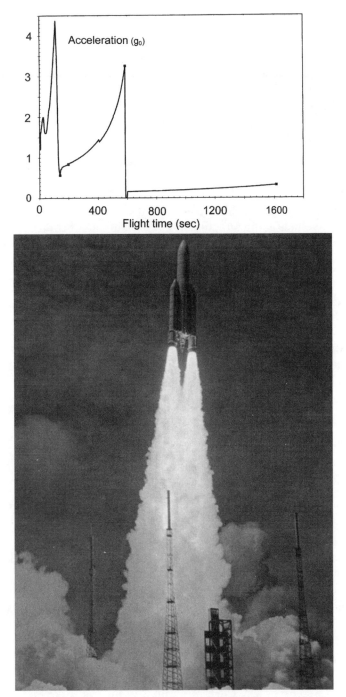

Figure 5.9: The acceleration of an Ariane 5 launch vehicle, expressed in *g*'s, as a function of flight time. (Images courtesy of Arianespace.)

airplanes, the exploitation of space as a potential holiday destination also becomes a reality.

However, the principal obstacles to achieving this vision of the future are the technical challenges that it poses, which go hand-in-hand with the large costs that would be incurred in the development of such a new generation of launch vehicles.

What are the technical challenges? The sort of launcher we are envisioning is referred to as a *single-stage-to-orbit* (SSTO) vehicle, and if you do the calculations to see if we can reach orbit with this type of vehicle, you find that it is just beyond our reach in terms of our current rocket technology. It almost feels like God created the planet, in terms of size and gravity field, so that it would be just that little bit too difficult to reach orbit with a SSTO vehicle using our current level of launcher technology. However, this paranoiac notion is tempered by the fact that this situation only applies now—to our capability at the turn of the 21st century. As time goes on, technology will develop, and it seems likely that the SSTO launcher will fly sometime in the next few decades, probably under the banner of a military research program.

What do we need to do to inject a useful payload mass into orbit using a SSTO vehicle? There are a number of approaches to the problem that can be taken, and a successful solution would probably combine all of them. The principal measures that can be taken include the following:

- An improvement in vehicle structural efficiency
- The enhancement of the process of optimization of the ascent trajectory to reduce the level of gravity and drag losses incurred
- The redesign and improvement of the performance of launcher's propulsion system

We discuss these measures in terms of the level of technical challenge they pose.

Structural efficiency: A conventional launcher typically carries hundreds of metric tonnes of liquid propellant and oxidizer, contained within large tanks distributed throughout the various stages of the vehicle. The safe containment of this mass of liquid in a harsh launch environment (of high acceleration and vibration levels) is a challenge to the structural designer of the launcher. The crux of this challenge is to manage this safe containment with the minimum of structural mass, so as not to compromise the launcher's ability to inject a good payload mass into orbit. A measure of the structural efficiency of a launcher is the ratio of the mass of its structure to the mass of fuel on board. Currently values of this ratio are on the order of 0.1, which means that for every 100 metric tonnes of fuel on board, we need typically 10 metric tonnes of structure to safely carry it. The challenge for structural designers and material scientists is to reduce the value of this ratio—equivalent to improving structural efficiency—in order that the mass saving in launcher structure can be invested in increasing the mass of the spacecraft payload.

Trajectory enhancement: The process of optimization of the ascent trajectory would probably mean adopting a horizontal runway takeoff and a fairly gentle climb to orbit (to reduce gravity loss). The vehicle would then spend more time in the lower, denser part of the atmosphere, potentially causing an increase in drag loss. To overcome this, significant technical effort would have to be invested in the optimization of the shape of the launcher to make it more aerodynamic over the prescribed flight path. The objective would be to ensure that overall losses are reduced

Propulsion system: The improvement of the performance of the launcher's propulsion system is probably the most difficult technical challenge in the development of a SSTO vehicle. With a conventional launcher system, the oxygen needed to burn the propellant in the rocket engines is carried, usually in liquid form, in large massive tanks. The overall performance of the launch vehicle can be enhanced significantly if we can devise a way of reducing this mass by extracting the required oxygen from the atmosphere. This is, after all, how a conventional aircraft operates, burning its fuel using the oxygen coming in through the engine intakes. This type of propulsion is referred to as *air-breathing*. There are some difficulties here, though, perhaps the most obvious one being that as the launcher climbs to near-orbital altitudes there is no usable atmosphere left from which to extract the oxygen. Another issue is perhaps a little more subtle, and has to do with operating air-breathing propulsion systems at high speeds. To understand this, we need to think about fast aircraft and the jet engines they use to sustain high-speed flight. Usually people use the word *supersonic* to suggest high speed, but this actually means that the aircraft is traveling at a speed in excess of the speed of sound. At sea level, the speed of sound is around 340 m/sec (1115 feet/sec), which is about 760 mph. An airplane traveling at this speed does seem fast, but it is actually traveling quite slowly compared to a launch vehicle, which needs to reach speeds of 8 km/sec (5 miles/sec) to reach a LEO, which is of the order of 18,000 mph.

Another way of addressing how fast an aircraft travels is to compare its speed with the local speed of sound using a *Mach number*. An aircraft moving at the speed of sound is said to be traveling at Mach 1, and one moving at, say, three times the sound speed, at Mach 3. Thinking in this way, *ramjet*-powered airplanes can reach speeds up to about Mach 5, which is a quite impressive 3500 to 4000 mph. At such high speeds, the air is rammed into the intake, which compresses the air sufficiently to be able to dispense with much of the mechanical complexity that you normally find in jet engines. Consequently, the ramjet is a relatively simple device—effectively a tube with an intake at one end, an exhaust nozzle at the other, and a combustion section in the middle. The inflowing air is mixed with fuel (for example, kerosene) and ignited. The pressure produced by the high-speed flow into the intake compresses the air and fuel mixture, and effectively directs the explosively ignited gas out of the exhaust nozzle to produce thrust.

One critical attribute of the ramjet is that the intake air flow has to be managed to reduce its speed to a subsonic level in the combustor, in order to ensure that combustion takes place. If it were otherwise, the air-fuel mixture would not be there long enough for the burning of the fuel to take place. The reason why this is critical is that the process of slowing the incoming air actually produces a kind of drag force on the engine, which is why the maximum operating speed of a ramjet is limited to about Mach 5. Given that we require higher speed operation for an air-breathing launcher propulsion system, this factor seems to be a bit of a problem.

To attempt to overcome it, an air-breathing propulsion system called a supersonic combustion ramjet, also known as a *scramjet* for short, has been proposed to potentially increase flight speeds up to around Mach 15. As the name implies, the scramjet is similar to the ramjet, but combustion takes place in the air-fuel mix while it is flowing at supersonic speeds within the engine. As a result, some of the limitations of the ramjet are overcome, but further technical challenges are posed, not the least of which is the problem implied above about sustaining the engine combustion in such a high-speed flow. Another issue with a scramjet-powered launcher is that the entire underside shape of such a vehicle needs to be designed and optimized as part of the propulsion system. The underside forebody becomes part of the engine intake system, ensuring high-speed flow into the engine, and the underside aft section becomes part of the engine exhaust jet designed to maximize the resulting thrust.

As if all this wasn't difficult enough, another challenge posed by operating at such high speeds in the atmosphere is the heat generated in the launcher's structure caused by *atmospheric friction*. The vehicle's forebody and the leading edges of the wings will reach very high temperatures, leading to a requirement to develop appropriate cooling techniques and materials, so that the vehicle does not fall apart due to this extreme heating.

Recently, experimental flight test programs have been established, by both civilian and military agencies, to attempt to demonstrate high-speed flight using scramjet propulsion, and to look at the challenges of the hypersonic

Figure 5.10: Artist's impression of a X-43 flight vehicle. (Image courtesy of NASA.)

aerodynamic design of such vehicles. Figure 5.10 shows an artist's impression of a scramjet-powered vehicle in flight, in this case from NASA's X-43 experimental aircraft program. At the time of this writing, however, these programs have had limited success, demonstrating scramjet-powered flight up to speeds of around Mach 10, with powered flight sustained for only a short period, of the order of tens of seconds.

Regarding the fully reusable, single-stage-to-orbit launch vehicle with aircraft-like operations, we can now see that the vehicle propulsion has to operate in a variety of different ways in order to accommodate the ground-to-orbit *flight envelope*. The takeoff from an airport runway would require the use of conventional jet engines, to take the speed to about Mach 2 or 3 when ramjet operation becomes effective. At around Mach 5, the engines would need to switch operation to scramjets, taking the launcher to around Mach 15 and to an altitude where the air is too thin for continued air-breathing operation. The last boost to orbit speed and height would require the engines to operate as rockets. To achieve this *combined-cycle* operation for the propulsion system, while limiting the overall mass of the engines, poses difficult technical challenges—so much so that many rocket scientists have expressed doubts that a single-stage-to-orbit manned vehicle, taking off from a conventional runway, will ever be achievable. But then perhaps this is an overly pessimistic view. After all, commercial air transport was a similar pipe dream a century ago.

Something About Environment

TO discuss how spacecraft are designed, we need to know a little bit about the environment in which they operate. The design of just about every machine built by engineers is influenced by its operating environment. For example, the design of an automobile is influenced by a number of factors, such as vehicle robustness, reliability, safety, the minimization of fuel consumption, reducing aerodynamic drag, and so on. These requirements are governed by the *environment* in which the vehicle will operate, that is, the conditions it will encounter on urban roads and highways. So, as you drive to work you can see the outcome of this environment-driven design process. Given that the design requirements for each vehicle are similar in terms of the encountered conditions, the computerized design process used by auto manufacturers these days results in all the cars (of the same vintage) looking similar to one another, apart from minor cosmetic tweaks!

As the environmental aspects become more dominant, such as in the case of an airplane, it becomes even more difficult to distinguish manufacturers, with only the connoisseurs being able to tell an Airbus from a Boeing.

In spacecraft, we have already seen in Chapter 5 how the flight environment of the launcher influences the spacecraft's design. In this case, the launch environment, which entails high levels of acceleration, vibration, and noise, governs how the spacecraft structure is designed. The designer has to ensure that the spacecraft survives the few minutes' ride from launch pad to orbit (we will return to this issue again in Chapter 9).

From the above discussion, it is easy to see that once the spacecraft has reached orbit, the natural space environment it finds there will also influence the spacecraft's design, and the designers need to know about this environment and take it into account as the design evolves. The characteristics of the space environment have been investigated by space scientists over many decades, and this information has been grasped enthusiastically by the engineers, thus enhancing the performance of spacecraft in the space environment. This environment has many different aspects, but the ones most influential in spacecraft design are *microgravity*

G. Swinerd, *How Spacecraft Fly: Spaceflight Without Formulae,*
DOI: 10.1007/978-0-387-76572-3_6, © Praxis Publishing, Ltd. 2008

(or weightlessness), *vacuum*, *radiation*, and *space debris*. This chapter briefly discusses each of these from the point of view of both the scientist and the engineer. The former is driven by curiosity to know the nature of what's out there, whereas the latter is interested in understanding the impact the environment has on the design of the spacecraft.

Hostile or Friendly?

As far as humans and our machines are concerned, there are some aspects of the space environment that are definitely hostile, and these include vacuum, high-energy electromagnetic and particle radiation, and space debris. However, there are other aspects that could be considered friendly: there is effectively no gravity, the environment is (mostly) clean, and we need not worry about factors that cause the erosion and deterioration that we find down here on the surface of Earth, such as wind and rain. As you might expect, it is the hostile elements that have most effect on how the spacecraft is designed, and these will be the focus of most of what follows. Interestingly, the Sun has a major effect on all of the hostile elements, and we need to say a few things about its dominant role in governing the space environment.

The Sun Rules OK!

Where does the Sun's energy come from?

The region of space surrounding the Sun is not called the solar system simply because of the gravitational grip that the Sun exercises over its attendant planets. It is also because of the total dominance it has in governing the space environment from the inner planets out to a distance on the order of 100 to 200 astronomical units (AU). This outer boundary of the solar system, called the *heliopause,* is the place where the Sun's influence ceases and the interstellar medium—the stuff between the stars in our galaxy, the Milky Way—begins.

The Sun, our star, is something we tend to take for granted. We never question that it will rise each day to illuminate our daily routine. We never give a second glance to the ever-present source of beautiful light and warmth that makes for a perfect summer's day. This rather laid-back attitude is encouraged perhaps by the Sun's small apparent size; it subtends an angle of only half a degree on the sky. However, this apparent size hides its true scale. The Sun is an object of about 1,400,000 km (870,000 miles) in diameter with a mass some 330,000 times that of our own planet! If we did stop to reflect for a moment, it is a rather sobering thought that we are living only 150 million kilometers (93 million miles) away from a star! Fortunately for us, it is a rather stable star, its output being more or less constant over the last few billion years, and the astrophysicists tell us that it will stay that way for a few more billions years to come. This stability is derived from a long-term balance between the energy source within the Sun tending to blow it apart and the force of gravitation tending to hold it together.

I find it amazing that we did not understand the source of the Sun's energy until as recently as the 1930s, when physicists began to uncover the mysteries of *nuclear fusion*. As the name suggests, this is the process of fusing atoms together to form heavier atoms. The basic energy source that powers the Sun is the fusing together of hydrogen atoms to make helium atoms, and this involves the release of nuclear energy. The destructive capability of the hydrogen bomb is also frightening testimony to the power of nuclear fusion. Some years ago many drivers displayed a green bumper sticker saying, "Nuclear energy—no thanks!" accompanied by a smiley Sun. Ironically, the antinuclear campaign's logo of the sun represented the largest source of nuclear energy in the neighbourhood!

How does nuclear fusion work? To begin, there are 92 different kinds of naturally occurring atoms, or *elements*. The *Periodic Table* lists these naturally occurring elements, starting at number 1, hydrogen, and ending with number 92, uranium. The currently accepted model of an atom is that it has a tiny, compact nucleus at its center composed of protons and neutrons, and this is surrounded by a cloud of orbiting electrons. The protons and neutrons are subatomic particles having about the same mass as each other, of the order of 0.000 000 000 000 000 000 000 000 001 of a kilogram, while the electrons are much smaller in mass (by a factor of around 2000). The protons each carry a positive electric charge and the electrons a negative one, while the neutrons are electrically neutral. This electric charge is the same as the static electricity that can sometimes build up on your clothing. You certainly get to know it's there if you touch a radiator, and the static charge gives you a mild electric shock as it dissipates to Earth. As the discussion above suggests, electric charges come in two varieties: positive and negative. We find that two like charges—two positives or two negatives—exert a force that repel each other, while a negative and a positive charge attract one another. The strength of this electric force between charges behaves like gravity in that it is governed by an inverse square law (see Chapter 1).

The numerical position of each element in the Periodic Table depends on the number of protons in the nucleus, so we have the simplest and lightest element hydrogen at number 1, consisting of a nucleus with one proton. At number 2 we have helium, with two protons (and two neutrons) in the nucleus, and so on up to uranium with 92 protons in the nucleus. To help visualize this, Figure 6.1 shows a hydrogen and a helium atom where the various particles are represented as billiard balls. Of course, the particles are not really like billiard balls, but instead have all sorts of weird properties that we need not discuss (if you are interested in knowing more, then I would suggest you find a popular book on *quantum mechanics*, which is the physics of the small world of subatomic particles). The nature of the subatomic world can be summed up by saying that we don't have any idea, for example, what an electron is! I find it remarkable that we can build a global consumer industry—the electronics business—on the basis of this fundamental ignorance. The saving grace, of course, is that our current theories are good at predicting *how* an electron behaves, so we don't really need to know *what* it is to build a television set or a computer games console.

We can complicate things a little by noting that atoms of a particular element can also exist in several forms, called *isotopes*, with different numbers of neutrons. For example, an atom with one proton and one neutron in the nucleus is an isotope of hydrogen called deuterium. One proton and two neutrons gives another isotope of hydrogen called tritium. These isotopes are also illustrated in Figure 6.1. The other puzzling thing about atoms with more than one proton in the nucleus is why the positively charged protons do not repel each other, and cause the nucleus to fly apart. The answer is that the physicists have discovered another force, the *strong nuclear force*, that binds the nucleus together. We have now come across three types of force so far in this book: gravity, electromagnetism (which includes the electric force), and now the strong nuclear force. So far, scientists have discovered only four fundamental forces in nature. The strength of the strong nuclear force exceeds that of the electric force, but it is

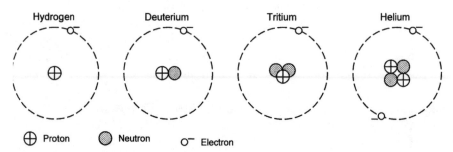

Figure 6.1: An illustration (not to scale!) of the hydrogen atom and the helium atom. The first two isotopes of hydrogen – deuterium and tritium – are also shown.

a very short-range force and so only acts, more or less, when the protons and neutrons come into contact with each other in a nucleus. The reason why there are only 92 *naturally* occurring elements is that when you get to number 93 (neptunium), it has 93 positively charged protons in its nucleus, and the repulsive electric force is just big enough to overcome the strong nuclear force, causing the nucleus to break up. At the time of this writing, scientists have created elements with up to 117 protons in the nucleus, but these heavy elements are unstable and don't hang around for long.

Now we can return to the process of nuclear fusion, which powers the Sun. The basic mechanism is to fuse together atoms or isotopes of hydrogen to form helium, and this results in the release of nuclear energy. The problem with fusion is overcoming the repulsive electric force that the protons have, and getting them close enough so that the strong nuclear force can grab hold of them and squeeze them together as a nucleus. Fortunately, the conditions in the core of the Sun are ideal for this to happen. The density is extremely high so that the protons are already very close together. The temperature is also extreme, on the order of 15 million degrees Celsius, which is such a large number as to make its meaning difficult to grasp. But the consequences for the protons is that, at these temperatures, they have high energies, and are rushing about at high speeds. This combination of the density and energy of the protons means that they can overcome their mutual electric repulsion, and get close enough for the strong nuclear force to bind them together. Thus we can form the nucleus of heavier atoms from light ones.

But where does the nuclear energy come from? In Chapter 1, we discussed what Einstein did for us. Another thing he gave us is an understanding that energy and mass are essentially different forms of the same thing. This he summarized in his famous equation $E = mc^2$, which basically says that mass m can be converted into energy E and vice versa (where c = 300,000,000 meters per second is the speed of light). I know this book isn't supposed to have any equations in it, but this one is so well known that it has become a part of our culture. It pops up in the titles of books and television programs. The thing to note about a helium nucleus is that, remarkably, it weighs less than the two protons and two neutrons (Figure 6.1) that compose it. Some of the mass has been used up in the energy associated with the action of the strong nuclear force in binding the helium nucleus together. To be precise, the helium nucleus has a mass just 99.3% of the mass of its parts. When the protons and neutron fuse together to form a helium atom, 0.7% of their mass is converted to pure energy in a way described by Einstein's famous equation.

We can do a simple sum to calculate how much of the Sun's mass is being converted every second into energy by the process of nuclear fusion. If we go out into the garden and present an area of one square meter to the Sun, the solar power falling on that surface is roughly 1.4 kilowatts (neglecting any losses that may occur due to passage through the atmosphere). If we now multiply this power by the number of square meters on a sphere the size of the Earth's orbit around the Sun, we can calculate the total power radiated by the Sun. Using Einstein's equation, we can then estimate the amount of mass being converted to energy each second. This turns out to be a staggering 4¼ million metric tonnes! This may sound like rather a lot of mass loss for the Sun to sustain, but the Sun can easily keep this up for many billions of years without it making much of a dent in the total mass. In fact, over the last 4½ billion years or so of the Sun's history, during which it has been shining at more or less a steady rate, about 100 Earth masses have been converted into pure radiate energy! Although this is a rather amazing statistic, nevertheless it represents only a tiny fraction of the mass available for nuclear energy production in the Sun.

Our very existence here on Earth is dependent on the Sun's stability, which is maintained by the Sun's massive gravitational field containing the awesome power of this nuclear furnace at its core.

The Sun's Output

With all this activity going on in the Sun, how does it affect us and our Earth-orbiting spacecraft from a distance of 150 million kilometers away? The main output from the Sun is radiation, and this comes in two varieties: electromagnetic (EM) radiation and particle radiation.

Light is one form of *electromagnetic radiation*. All forms of EM radiation travel at the speed of light (see above), and are distinguished from each other by the wavelength of the radiation. Figure 6.2 shows different types of EM radiation, from short wavelength *gamma rays* to long wavelength *radio*

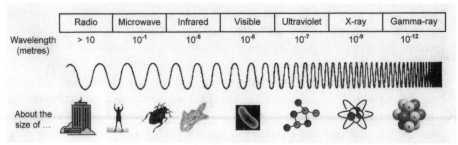

Figure 6.2: Visible light is just one part of the electromagnetic spectrum. Different types of EM radiation are distinguished by wavelength, from short wavelength gamma rays to long wavelength radio waves. The wavelength is about the size of the items pictured at the bottom of the diagram (the object that looks a bit like a peanut – about the size of the visible light wavelength – is supposed to be a small bacterium).

waves. The wavelength of each type of radiation is given in meters, using a shorthand notation used by scientists. For example, 10^{-1} m means 0.1 m, and 10^{-6} m means 0.000 001 m. Note that the number after the minus sign indicates how many digits there are after the decimal point. For example, gamma radiation has a tiny wavelength of 10^{-12} m = 0.000 000 000 001 m, about the size of an atom's nucleus.

The visible part of the spectrum—light—has a wavelength ranging from about 0.4 μm (violet) to 0.8 μm (red), where μm stands for micrometer (sometimes called a micron), which is a millionth of a meter = 10^{-6} m = 0.000 001 m. Slightly shorter wavelengths take us into the *ultraviolet* region (which can cause sunburn on a sunny day), and the slightly longer wavelengths take us into the *infrared* region (which is basically heat radiation, such as that which you feel on your face in front of a glowing open fire).

The Sun emits EM radiation across the entire spectrum, but the peak of its output is at a wavelength of about 0.5 μm, which is in the yellow part of the visible light spectrum. Evolutionary theory says that this is why our eyes are most sensitive in this part of the spectrum, the eye having evolved in an environment dominated by sunlight. The harmful emissions of gamma and X-rays from the Sun are fortunately (relatively) small. This short wavelength radiation is particularly hazardous to humans. It is, for example, one cause of radiation sickness after exposure to a nuclear bomb detonation. There is also a significant intensity of ultraviolet radiation, but fortunately we are protected from most of this at ground level by the famous *ozone layer*. We are now aware, however, that dangerous holes are being punched in this protective shield by the inadvisable use of certain artificial chemicals. Above the atmosphere, spacecraft are of course exposed to the full range of the EM radiation spectrum from the Sun.

The other form of radiation from the Sun, which can damage orbiting spacecraft and people, comes as a stream of energetic (high speed) subatomic particles called the *solar wind*. The source of this is the violent eruptions that take place on the Sun's searingly hot surface (at a temperature of around 6000°C), and in its atmosphere. Material is flung into space mostly in the form of protons and electrons, but also in the form of the nuclei of atoms stripped of their electrons—called *ions*. By the time it reaches Earth, this steady stream of solar wind has a density of a few tens of particles per cubic centimeter and is traveling at a speed typically between 300 and 1000 km per second (670,000 to 2,240,000 miles per hour). Despite the relatively low density of this stream of ions, it does have a significant effect on our planet, and in particular on the Earth's magnetic field. The Earth has its own magnetic field, which looks a bit like that of a bar magnet. Science teachers often demonstrate magnetic fields by sprinkling iron filings onto a sheet of paper, which in turn is placed over a bar magnet. By jiggling the paper, the iron filings outline the shape of the bar's magnetic field, to reveal a pattern like that shown in Figure 6.3a. This classic shape of Earth's magnetic field is referred to as a *magnetic dipole*. However, the solar wind also carries a magnetic field, and when this encounters Earth's field, the classic dipole shape is disturbed considerably.

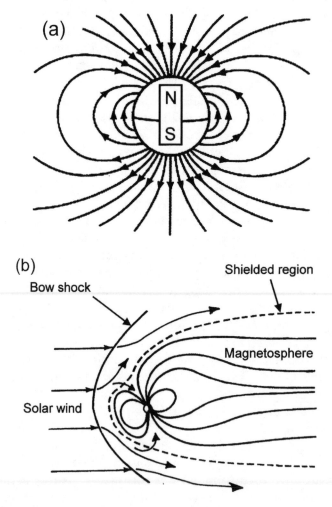

Figure 6.3: (a) The shape of the magnetic field of the Earth resembles that of a bar magnet. (b) Earth's magnetic field is changed by its interaction with the solar wind. Generally the solar wind particle radiation is deflected by Earth's protective magnetic field, although some charged particles are trapped in the magnetosphere, and some reach Earth over the polar regions causing auroral displays.

This *solar-terrestrial interaction* between the solar wind and the Earth's magnetic field is complex, and acquiring an understanding of it has stretched the intellect and imagination of many talented scientists over a number of decades of research. To gain some insight ourselves, and to appreciate why it is important, we need to think about some basic aspects of electricity and magnetism. The first thing to note is that an electric current in a wire produces a magnetic field. This idea has been around for a long time, being first demonstrated in 1820. That this is true can be easily seen by placing a compass needle near a wire carrying electricity. Normally the needle would align itself along Earth's magnetic field and point north, but the electrical current produces its own magnetic field that disturbs the needle so that it no longer does so. It is a simple job to repeat this historic experiment with a compass, a length of insulated wire with the insulation stripped off the ends, and a battery. To make it work, we need direct current (DC)—a flow of electrons in the wire in one direction only, which is provided by the battery. A nearly expired AA battery from a portable CD player or similar device would do fine, as opposed to a fresh one. (We are going to effectively short the battery with the wire, so the battery will probably be no good afterward!) The setup is illustrated in Figure 6.4. The wire is positioned as close as possible above the compass needle, such that the wire is parallel to the north-pointing compass needle. It is sometimes easier to tape one end of the wire onto one of the battery terminals to ensure good electrical contact. If we gently stroke the other end of the wire on the second battery terminal, completing the circuit, we can easily see the current's magnetic field kicking the compass needle, tending to cause it to point in a direction at right angles to the wire.

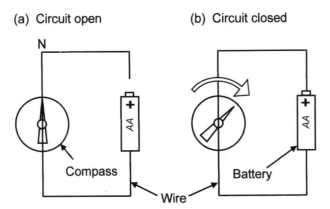

Figure 6.4: A simple setup to demonstrate that a current produces a magnetic field.

It is easy to see a similarity between the solar wind and the current in the wire: both are rapidly moving streams of charged particles. In the wire, the current is made up of a flow of charged electrons, whereas in the solar wind the current is generated by a stream of ionized (charged) particles emanating from the Sun. However, the point is that the solar wind carries its own magnetic field, and when this hits Earth's magnetic field, the classic dipole shape is squashed on the Sunward side, and stretched in the down-Sun direction into a shape something like that shown in Figure 6.3b. This region filled by Earth's magnetic field is referred to as the *magnetosphere*.

The fact that Earth has a magnetic field is a bit of a saving grace in itself, as it prevents the damaging effects of the solar wind from reaching Earth's surface directly. Instead, the solar wind's stream of particle radiation passes through the "bow shock," which slows down the flow, before it is deflected around the magnetosphere. The explanation of this shock front is a bit technical, but basically it is similar to the shock wave in front of a supersonic aircraft. Although the major part of the radiation is diverted, some particles are trapped in Earth's field, and some penetrate Earth's defenses and are funneled down onto the north and south magnetic poles, producing the spectacular manifestation of the Northern and Southern Lights. Otherwise know as *aurora*, these subtle, colorful, and dynamic displays of glowing lights seen in the night sky at high latitudes can arouse a sense of awe in even the most jaded and uninterested of individuals. The glow in the air is caused by exactly the same mechanism as the light coming from a neon strip light; when we switch it on, a flow of charged particles (electrons in this case) is passed through the neon gas, causing the atoms of neon to glow with a characteristic "white" color. Similarly, when the charged particles from the solar wind race down through the atmosphere over the polar regions, the air glows with colors characteristic of the different gases, mainly oxygen and nitrogen, thus producing the auroral display. This intense flux of charged particles also dumps vast amounts of energy into the atmosphere, resulting in an increase in atmospheric temperature of the order of hundreds of degrees Celsius at high altitudes.

The intensity of the Sun's output, both EM radiation and solar wind, varies over an 11-year period called the *solar cycle*. Roughly every 11 years the output goes from a maximum, through a minimum, and back to a maximum again, with the last peak (at the time of this writing) occurring approximately in the year 2001.

At times of solar maximum, the disturbing effects of our nearest star become even more vigorous. The frequency and violence of outbursts on the Sun's surface, referred to as *solar flares*, increase. These hurl billions of metric tonnes of ionized material into interplanetary space. As a consequence, energetic "wodges" of solar wind move outward from the Sun at high speed, and if Earth just happens to be in the wrong place at the wrong time, it will be enveloped in this cloud of energetic charged particles—an event sometimes referred to as a *solar storm*. A picture of such a wodge— more correctly referred to as a *coronal mass ejection* (CME)—is shown in Figure 6.5. This lovely image was acquired by one of the Stereo spacecraft in January 2007. An intense pulse of solar wind is shown on the right-hand side of the picture, with Venus at the bottom left and Mercury at the bottom right. During solar maximum, the frequency and intensity of auroral displays increase, the upper atmosphere heats up and expands, and orbiting satellites can receive damaging and sometimes fatal doses of particle radiation. Electrical power grids on the ground can also come under attack, due to the interaction between the Earth's and the solar wind's magnetic fields. During a solar storm, the magnetic field of the solar wind is relatively intense and time-varying, and this buffets the Earth's field, squashing and stretching it in response to the solar bombardment. At ground level the resulting movement of the magnetic field can induce surges of electrical current in long

Figure 6.5: A coronal mass ejection propagating across the inner solar system, with Venus and Mercury clearly seen in the lower part of the image. The Sun is just off the right-hand side of the picture. (Image courtesy of the National Aeronautics and Space Administration [NASA].)

conductors, such as power lines and pipelines, which can cause terrestrial power systems to overload and blackout. Notably, such an occurrence caused a massive blackout in the province of Quebec, Canada, during a solar maximum storm in 1989.

The fact that the movement of a magnetic field relative to a wire induces an electric current in the wire has also long been known. This principle of electromagnetic induction was first discovered by Michael Faraday in 1831, and was greeted at the time as an interesting curiosity. However, in the years following, this discovery led to the vast industrial application of electrical power generation that has transformed every aspect of our technological world. Put simply, modern power generation is achieved by huge generators that are rotated rapidly by some means, such as heating a fluid to drive a turbine, which in turn drives the generator. The heat source can be through the burning of coal or oil, or by the harnessing of nuclear power. But the point is that the generator rotates a huge harness of wire in a magnetic field, which induces an electrical current in the wire, thus producing electrical power to supply the national grid. Consequently, we have seen that, on the one hand, a moving stream of charged particles (an electrical current) can produce a magnetic field, and, on the other, the relative movement between a wire and a magnetic field can induce an electrical current in the wire. During the latter half of the 19th century, electromagnetic theory was developed, principally by the Scottish physicist James Clerk Maxwell, and we have come to understand these two effects as opposite sides of the same coin. The development of Maxwell's equations, which lay down the theoretical framework for electromagnetism, ranks as one of the greatest achievements of 19th century physics.

Returning to the nature of the Sun's output, and its total dominance over the solar system environment, we are rather fortunate to have a planet that provides an ozone shield to protect us from electromagnetic ultraviolet radiation, and a magnetic field to protect us from high energy particle radiation. Today, as I write, just happens to be one of those cold, clear, bright November mornings that seem so rare in British winters. The Sun, despite being low on the horizon, is absolutely brilliant, seemingly dominating the whole of the eastern hemisphere of the sky, in keeping with its awesome power and its influence over life on Earth and over the solar system in general. After reading this, my hope is that when you're on your way to work tomorrow, you too might look skyward and contemplate how remarkable is our companion star—in fact, if you reflect on it too much, it can be daunting that all that stuff is happening just a short distance away (in cosmic terms) from where we live!

The Impact of the Space Environment on Space-craft Design

Microgravity

One aspect of the space environment that is not affected by the Sun is microgravity. This term describes the state of *weightlessness* that we discussed in Chapter 2 (see Figure 2.2). As described earlier, once a spacecraft has reached its mission orbit, it is in a state of continuous free-fall so that it effectively encounters an extended period of weightlessness. Obviously, from the point of view of the design of the structure of the spacecraft, this is a gentle situation. It is interesting that, as a consequence, the structural design of the spacecraft is mostly driven by the few minutes' ride into orbit on a launch vehicle, and not by the 10 years or so of operational life on mission orbit. There may be the odd exception to this, such as highly maneuverable military spacecraft, but it does hold true in general for most civil scientific or commercial satellites.Microgravity is generally a rather friendly aspect of the space environment, but it can pose some intriguing problems for space engineers that sometimes require ingenious solutions. One of these, which we have already mentioned in Chapter 5, is how to persuade liquid rocket propellant to enter a rocket engine when it is in a weightless state in the fuel tanks. One common way to do this is to have a rubber diaphragm stretched across the middle of the fuel tank, with propellant on one side and a pressurized gas on the other. When the fuel outlet valve is opened, the diaphragm with the gas pressure behind it squeezes the fuel out of the tank and into the rocket engine.

Another puzzle is how to ground-test devices on the spacecraft that are intended to be deployed and operated in a microgravity environment. This is a common problem for spacecraft test engineers. To perform its mission, a spacecraft often has to deploy solar arrays, large antennas, and possibly other flexible pieces of equipment once it reaches its mission orbit. However, for this equipment to fit into the launcher, it has to be folded up and secured to the spacecraft in a compact and robust arrangement, to ensure its survival during the ride to orbit. Of course, this means that once in orbit, the various deployable items need to be unfurled into the operational arrangement, and this is often done using a variety of spring mechanisms. The problem for the engineers is how to test these deployment mechanisms on the ground (in a $1g$ environment) that are intended to be operated in weightlessness conditions on orbit. Although this is a complex problem that requires simulating zero gravity on the ground, it is often solved by simple means, such as hanging deployable items from wires to take their weight, or even attaching helium-filled balloons to counter their weight, while tests are carried out.

Vacuum

A pure vacuum—a volume devoid of any material whatsoever—is something not yet encountered by scientists. As we found in Chapter 3, even in Earth orbit at altitudes up to around 1000 km (620 miles) there is a residual atmospheric density, and even in the spaces between the planets in the solar system there is material, called the *interplanetary medium* (mainly emanating from the Sun, as we have seen). However, as far as people and spacecraft are concerned, the degree of vacuum is fairly academic. Over millions of years, we and our ancestors have adapted to an environment where every square centimeter of our bodies is exposed to an atmospheric force of about 10 Newtons, which translates into imperial units as the familiar sea-level pressure of about 15 pounds per square inch. If we remove this pressure by foolishly stepping out of a spacecraft without protection, then surprisingly tests show that we don't explode, or anything dramatic like that. The main problem is that there is no oxygen to breath, and consciousness is lost after a few tens of seconds, followed by death after a couple of minutes. Precisely what happens, and when, is not well known, as scientists are understandably reluctant to do too many experiments! For spacecraft, the effects of high vacuum are rather less spectacular, but nevertheless the spacecraft designer needs to know something about it to avoid using the wrong materials in the spacecraft's construction.

At about 800-km altitude in Earth orbit, the atmospheric pressure is tiny (of the order of 0.000 000 000 001 Newtons per square centimeter), and at these low pressures materials suffer an effect called *outgassing*. This is related to what happens to water when heated—the surface water molecules escape the body of the liquid, and if the process continues, all of the water will vaporize into gas. Similar things happen to metals in high vacuum, where the low pressure causes the surface atoms to outgas. For example, at temperatures of around 180°C, a surface composed of zinc will recede at a rate of around 1 mm per year. However, for a material like titanium—one much more commonly used in spacecraft construction—a temperature of 1250°C is required to achieve the same rate of recession. Thus, as long as the designer chooses the construction materials appropriately, outgassing will not be an issue as far as the strength of the structure is concerned. But sometimes there is a concern over the outgassing material contaminating the spacecraft's surfaces; for example, the performance of a space telescope may be compromised if outgassed material is deposited onto the system's optics.

A related problem for spacecraft is the effect of vacuum upon commonly used terrestrial lubricants. The highly volatile oil-based lubricants we use in our machines down here would outgas (or boil away) in no time at all in the vacuum of space, which has given rise to a whole new science of *space tribology*. To overcome this problem, engineers have had to develop solid lubricant coatings for use in spacecraft bearings and mechanisms. Interested readers can 'google' the term *molybdenum disulfide* which is commonly used as a solid lubricant.

The Effects of Earth's Atmosphere

As we saw in Chapter 3, the motion of spacecraft in low Earth orbit is affected by the atmosphere. The cause of this is air drag, which you may recall is a tiny force that acts in a direction opposite to the motion of the spacecraft. We saw how this takes energy out of the orbit, causing the spacecraft's height to decrease, with the ultimate prospect of a fiery reentry into the denser, lower atmosphere. Of course, there is no altitude where we can say that the atmosphere stops and the interplanetary medium begins. The density of the atmosphere falls steadily, from the breathable mix of oxygen and nitrogen at Earth's surface, to something approaching a vacuum at high altitude. However, we can measure the effects of air drag at altitudes up to about 1000 km (620 miles). Earlier we discussed the Sun's dominant influence on Earth's environment. The drag effects on a low Earth orbiting spacecraft are also significantly affected by the Sun. Over the 11-year solar cycle, the level of solar activity varies, resulting in peaks and troughs in its electromagnetic and particle radiation output. At times of solar maximum, intense EM radiation from the Sun, in particular in the ultraviolet part of the spectrum, causes the temperature of Earth's upper atmosphere to rise.

It's worth thinking for a moment about what we mean by temperature in this outermost layer of Earth's atmosphere, referred to as the *exosphere*. Given that the atmosphere is close to a vacuum at these high altitudes, if we tried to use a thermometer to measure the temperature, there would not be enough air around to give any sort of sensible reading. Instead, when we refer to temperature in such tenuous material, we usually talk about something called *kinetic temperature*. When the Sun's ultraviolet radiation heats up the atmosphere, it essentially "excites" the atmospheric particles and causes them to race around at a higher speed, thus increasing their kinetic temperature.

It is important to realize how tenuous the atmosphere is at high altitudes. For example, an atom of oxygen moving around at an altitude of, say, 600 km (370 miles) will not find another atom to bump into for about 300 km (185 miles), and at an altitude of 800 km (500 miles), this increases to over 1000 km (620 miles). The precise values of these numbers actually vary with the level of solar activity, but the main point is how thin the atmosphere is at these heights. Essentially the atoms and molecules that make up the upper atmosphere move around on *ballistic trajectories*, similar to how cannon-balls fly in a gravity field. This may be simplifying the matter a little, but it does help us understand how solar heating of the atmosphere causes its density to increase.

Putting it all together, then, at times of solar maximum the Sun's ultraviolet output increases, which in turn causes a rise in temperature of Earth's atmosphere. This causes the atmospheric atoms and molecules to rush around more rapidly, allowing them to reach higher altitudes in Earth's gravity field. At a particular height the numbers of atmospheric particles per cubic meter increase, producing a higher density. This effect on the atmosphere is significant; for example, the density of the atmosphere at, say, 600 km at high solar activity can be larger than that at low solar activity by more than a factor of 10. We are not just talking about increases of two or three times, but more than an order of magnitude.

In considering the perturbing effect on a spacecraft orbiting at a 600-km altitude, the drag acting on it is related directly to the atmospheric density, so drag at times of high solar activity can be more than a factor of 10 greater than when the Sun is quiet. This will have a significant effect on spacecraft operations, and the mission analysis team will have to take the solar cycle into account in planning the orbit control activities and in the estimation of how much rocket propellant will be required to compensate for these drag effects.

To show the huge effect that the solar cycle has on atmospheric temperature, Figure 6.6 is a chart of the variation in the average exospheric temperature (the temperature of the atmosphere at orbital altitudes) over the last five solar maxima—since the dawn of the space age. Most surprisingly, the upper atmosphere reaches kinetic temperatures of the order of 1000°C when the Sun's level of activity is high. The 11-year solar activity cycle can also be clearly seen, with the atmospheric temperature at solar maximum being something like 600°C above that at solar minimum. Another striking feature of Figure 6.6 is the spiky nature of the temperature profile at times of solar maximum. This spike is due to temperature variations caused by more short-lived events such as the solar storms mentioned above. During such storms, some of the solar wind particles from the Sun are focused down into the atmosphere at the north and south polar regions by Earth's magnetic field, causing auroral displays. This influx dumps massive amounts of energy into the atmosphere, which causes heating locally at the poles. Within a short period (of the order of hours), this heating propagates toward the

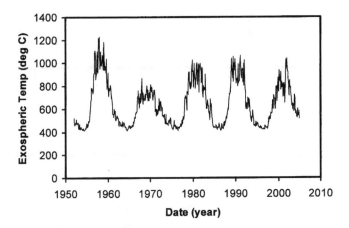

Figure 6.6: The variation of average exospheric temperature over the last five solar maxima. (Figure compiled from data supplied courtesy of Dr. Hugh Lewis, University of Southampton, UK.)

equator, again causing the atmosphere to expand. This produces an increase in atmospheric densities at orbital altitudes, giving a corresponding increase in drag perturbations on spacecraft orbits.

The atmosphere also has more direct effects on spacecraft and the materials used in their construction. Perhaps the best example of this is *atomic oxygen erosion*. Down on the Earth's surface we breathe molecular oxygen O_2, composed of two oxygen atoms chemically bonded together. We know that the oxygen we find down here gives us life, but it is also aggressive in forming oxides such as rust, which if left untreated over a period of years will have a damaging effect on our machines (bikes, cars, lawn mowers). As you go up to orbital altitudes, the atmosphere is no longer shielded from solar ultraviolet radiation. This causes the O_2 bond to be broken, so that at low Earth orbit (LEO) altitudes single oxygen atoms (referred to as atomic oxygen and denoted by the symbol O) wander around and become the dominant atmospheric constituent.

Atomic oxygen on orbit has a similar erosive character, not only arising from its chemical activity, but also because it hits spacecraft at around 8 km/ sec (5 miles/sec) (due to the vehicle's orbital speed through the atmosphere). The importance of this was first registered when most of the thermal blanket on a camera mounted on the Space Shuttle during its third mission in March 1982 disappeared due to the effects of atomic oxygen erosion. Thermal blankets are used extensively on spacecraft to insulate them from the heating effects of direct solar radiation (see next section), and often give the vehicle its characteristic appearance of being wrapped in gold or silver foil. The blanket is composed of multiple layers of a thin plastic film with a metallic coating, such as aluminium, silver, or gold, similar to the survival blankets handed out at the end of marathons to keep the runners warm (see Chapter 9). In addition to the thermal blanket, various other materials are also particularly prone to atomic oxygen attack, one example being silver, which is commonly used in the construction of solar panels (solar panels are used on spacecraft to convert sunlight into electrical power; see Chapter 9). Clearly, the spacecraft designer needs to be familiar with these environmental effects when choosing appropriate materials in the design.

Electromagnetic Radiation

Most of the energy in the Sun's electromagnetic spectrum is contained within wavelengths ranging from about 0.2 to 3 μm (see Figure 6.2), ranging from short wavelength ultraviolet radiation, through visible light, to longer wavelength infrared (heat) radiation. The most obvious effect of this radiation on an orbiting spacecraft is the thermal heating that it causes. For an Earth-orbiting spacecraft, the solar power falling on every square meter of surface presented to the Sun is about 1.4 kilowatts, so that the heat input to the spacecraft surfaces is substantial. By contrast, a spacecraft in a LEO usually enters Earth's shadow on each orbit, and when this happens the vehicle's surface temperature drops drastically. Management of this thermal cycling is a critical job to be done by the thermal control subsystem engineer (see Chapter 9) to ensure that the equipment inside the spacecraft does not suffer a damaging level of temperature variation.

Other more direct impacts on design come from the damaging effects of the short wavelength solar ultraviolet radiation. Down here at the Earth's surface, as we have seen, its harsh effects are softened by the protective ozone layer, but an orbiting spacecraft is exposed long-term to its erosive effects. The chemical structure of paints and thermal blanket material can be modified by ultraviolet radiation, causing them to become brittle and flaky. In general, the spacecraft's surfaces will erode, due to ultraviolet radiation, in such a way that the amount of the Sun's heat that they absorb will increase over time, which again is of concern to the thermal control engineer (see Chapter 9).

Trapped Particle Radiation

As we saw earlier, some of the solar wind particles emanating from the Sun, generally charged particles such as electrons, protons, and atomic nuclei stripped of their attendant electrons, penetrate the protective shield of Earth's magnetic field. Some are focused by the field into the atmosphere above the north and south polar regions, causing auroral displays. Others are trapped by the magnetic field, producing radiation belts that pose a hazard to people and spacecraft alike. These belts are called the *Van Allen belts*, after their discoverer James Van Allen, who was the first to confirm their existence using data from the Explorer 1 and 3 satellites in 1958.

These charged particles, once trapped by Earth's magnetic field, move rapidly in a particular way governed by the field, to give the Van Allen belts their characteristic shape of giant doughnuts stretched around Earth's equator (see Figure 6.8).

To understand how this structure comes about, we need to consider how charged particles move in a magnetic field. Put simply, they tend to gyrate about the magnetic field lines; their paths through space echo a shape a bit like a corkscrew, as shown in Figure 6.7a. If we also recall the shape of Earth's magnetic field, which is similar to that of a bar magnet (see Figure 6.3a), then the particles travel paths like that shown in Figure 6.7b. A trapped

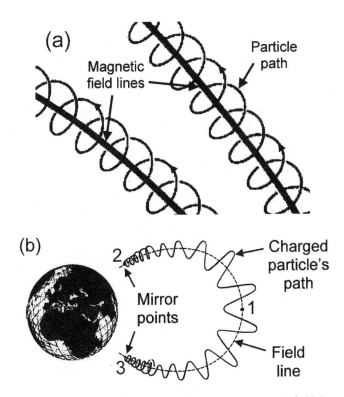

Figure 6.7: (a) Charged particles gyrate around the magnetic field lines, traveling paths that resemble a corkscrew. (b) Charged particle radiation is trapped in Earth's magnetic field, bouncing rapidly between mirror points 2 and 3, producing the Van Allen radiation belts.

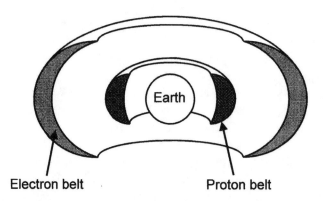

Electron belt   Proton belt

Figure 6.8: A slice through Earth's trapped radiation belts reveal a shape that echoes the magnetic field. The regions of maximum intensity of the proton and electron belts are shown.

particle moving north across the equator at point 1 will corkscrew about the field line toward point 2. As it approaches Earth at point 2, the magnetic field strength increases and the field lines converge, producing a *mirror point*, which causes the particle to bounce back along the field line toward point 1 again. At the other end of its journey, point 3 also acts as a mirror point, which reflects the particle back toward its starting point again. Thus our trapped particle is destined to bounce back and forth between points 2 and 3 indefinitely, unless it leaks away into space or is captured by Earth's tenuous upper atmosphere. This gives the characteristic shape of the Van Allen belts as shown in Figure 6.8. These are regions where spacecraft and people should not linger, due to the high density of energetic particle radiation, composed mainly of high-energy electrons and protons. The maximum intensity of electron radiation occurs at an altitude of approximately 27,500 km (17,000 miles), and the greatest intensity of the more damaging proton radiation occurs at around a height of 4500 km (2800 miles).

We would not contemplate orbiting a manned space station at these sorts of heights, as the consequences for the crew would be serious. However, we do know that exposure of humans to the radiation belts for a short spell, although not desirable, is tolerable; for example, the Apollo astronauts had to fly through the belts on their way to the moon and on their return.

Long-term exposure of unmanned spacecraft to the radiation belts is also damaging, mainly causing degradation of electronic and electrical power systems. Deploying a satellite in a circular orbit at the heart of the Van Allen proton belt would be foolhardy, but as mentioned in Chapter 2, many spacecraft occupy large elliptical orbits and fly through the radiation belts on each orbit revolution. The main problem for these vehicles is that their solar panels suffer radiation damage, which causes the amount of power they produce from sunlight to decrease with time. A spacecraft in this type of orbit for many years may suffer a power loss up to 50% of the solar panel's original output. However, the power subsystem engineer is able to predict the likely deterioration for the particular type of orbit flown, and make due allowance in the spacecraft's design.

Other electronic components onboard are also subject to radiation damage, but unlike solar panels, they can be shielded to some degree from the energetic particles by increasing the thickness of the walls of the metal boxes (typically made of aluminium) in which they are usually mounted. However, this needs to be done carefully as it will increase the spacecraft's mass, and as we have seen in Chapter 5, an increase in mass means a larger, more expensive launch vehicle. Another way of providing radiation protection, which goes some way toward solving this mass-growth problem, is to place radiation-sensitive components sensibly within the spacecraft so that they are shielded by less sensitive adjacent equipment.

As well as these *total dose effects*, the operation of onboard electronic equipment is also disrupted by *single-event upsets* (SEUs). These are temporary effects caused by the passage of a single high-speed particle through a computer processor, for example, producing random bit flips in onboard software, that is, switching a 0 bit to a 1 bit in a computer program, which can have undesirable and unpredictable results onboard! To help overcome this, spacecraft computer programs include error correction codes, which continuously and routinely check the onboard computer memory. Another type of problem caused by particle radiation, a *single event burnout* (SEB), is much more serious as it can cause permanent damage to the spacecraft's electronic systems. In this case a single energetic particle can kick off a runaway current in an electronic component, causing the device to burn out. To overcome this, the designer can build in more shielding, or choose to protect the device with current-sensing and -limiting circuitry, but again there is a balance to be struck to prevent mass growth in the spacecraft design.

In this brief tour of how environmental effects influence spacecraft design, the final topic involves impacting particles again, but this time somewhat larger ones than the subatomic particles we find in the Van Allen belts.

Space Debris

Space debris comes in two varieties: natural and artificial. *Natural space debris* consists of the meteoroids that Earth encounters on its orbital journey around the Sun. These meteoroids themselves are in orbits around the Sun, and they vary in size from a few meters in diameter to tiny specks of material smaller than a grain of sand. It is estimated that around 100 metric tonnes of such material rains down into Earth's atmosphere on a daily basis, although there is controversy about this estimate. Fortunately, most of it burns up harmlessly in the atmosphere, producing the fleeting flash of light that is associated with a shooting star. The vast majority of such meteoroids are at the tiny end of the size spectrum. However, they can still have significant energy, due to speeds on the order of a few tens of kilometers per second. Fortunately, the chances of an orbiting spacecraft encountering a large meteor (say a few centimeters across) are negligible, which is just as well as the consequences would be catastrophic. For example, the energy of a 2.5-cm (1-inch) rocky meteoroid traveling at 20 km/sec (12.4 miles/sec) is about the same as a 20-ton truck traveling at 110 km per hour (70 mph), and such a projectile would make short work of an orbiting satellite! But the probabilities are such that the main issue with natural debris is the peppering of the spacecraft's surfaces by high-velocity dust particles, producing a general degradation of thermal blanketed and painted surfaces.

The threat to satellites posed by *artificial debris*, on the other hand, is much greater, as once the debris size reaches a millimeter and above, the artificial debris objects in orbit begin to outnumber the natural debris objects. And this trend continues, so that the chances of an impact with a 10-cm (4-inch) chunk of artificial debris is much greater than the odds of encountering a meteoroid of similar size.

Artificial debris, as the name implies, are useless lumps of man-made material in space that have ended up in orbit as a by-product of launching spacecraft. Various agencies have proposed formal definitions of artificial space debris. The United Nations Committee on the Peaceful Uses of Outer Space states, "Space debris is defined as all man-made objects, including fragments and elements thereof, in Earth orbit or re-entering the atmosphere, that are non-functional." Space turns out to be just another arena of human activity that is being steadily polluted. However, the junk we leave in space is generally more dangerous than our terrestrial garbage as it is moving about at high speeds, posing a potentially lethal hazard to people and spacecraft in near-Earth orbits. Since the dawn of the space age in October 1957, when the first satellite Sputnik 1 was launched, there have been a total of 27,000 catalogued objects launched, with about 9000 catalogued objects still currently in orbit (at the time of this writing). Indeed, the upper stage of the launch vehicle that put Sputnik 1 into orbit also entered orbit, and became the first item of artificial space debris with a mass of about 4 metric tonnes. A *catalogued object* is any object large enough to be routinely tracked by a number of ground-based sensors to allow its orbit to be determined. Once the object's orbit is known, its details are placed in the U.S. Space Command catalogue, with its own unique catalogue number. The sensors used to do this job are mostly large radars, which once comprised the ballistic missile early warning system used during the Cold War. The sensitivity of these sensors is such that any object larger than about 10 cm (4 inches) in LEO and larger than around 1 m (3 feet) in geostationary Earth orbit (GEO) are tracked and catalogued.

Of the 9000 current catalogued objects, about 5% are operational spacecraft, but the majority have no useful function. Of this majority, many are large derelict upper stages of launch vehicles, which have accompanied their spacecraft payload into orbit. Others are smaller objects that are released into space during the process of launching spacecraft. About 40% of the total are fragments resulting from the accidental, explosive breakup of upper stages or spacecraft in orbit. In many cases the unintentional mixing of leftover propellant and oxidizer in a launcher upper stage has produced an explosive event that has torn the upper stage apart, producing several hundred new catalogued objects overnight!

A simple subtraction—27,000 minus 9000—gives us about 18,000 objects that have either flown away into interplanetary space or have fallen from orbit and reentered Earth's atmosphere. Most of these objects have indeed come back through the atmosphere, burning up harmlessly, although there are some famous instances when large pieces of spacecraft have reached the ground (such as the unwelcome arrival of parts of a nuclear reactor from Cosmos 954 over Canada in 1978, and Skylab's reentry over Australia in 1979). Figure 6.9 shows the number of catalogued objects in LEO over time from 1957 to 2001. There are about another 1000 or so objects in other orbital regions, making up our total number of approximately 9000. Figure 6.9 shows an almost continuously rising trend. However, interestingly we can see that this trend is interrupted, particularly in the early 1980s, the early 1990s, and around the year 2000, when the curve dips or flattens out. These periods correspond to times when the Sun's activity was at a maximum, producing atmospheric heating, with a resulting increase in atmospheric density (see The Effects of Earth's Atmosphere, above). This in turn produced a rise in drag on orbiting satellites, increasing the numbers of reentries into the atmosphere. So we can say that solar maximum is a time to put on the hard hats!

The distribution of catalogued objects in LEO with orbit height is shown in Figure 6.10. We see that there are few objects at low altitude, say, less than 500 km. This is because the drag perturbations on debris are relatively large (due to the higher atmospheric density at low altitudes), sweeping the debris

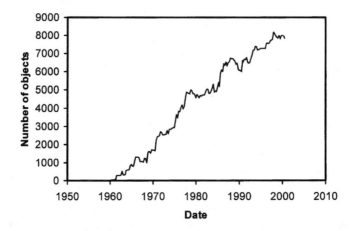

Figure 6.9: Numbers of objects greater than 10 cm (4 inches) in size in low Earth orbit (LEO) over time, from Sputnik 1 (October 1957) to 2001. (Figure compiled from data supplied courtesy of Dr. Hugh Lewis, University of Southampton, UK.)

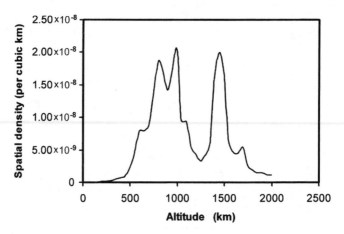

Figure 6.10: The current distribution of objects in LEO, greater than 10 cm (4 inches) in size, plotted against height. (Figure compiled from data supplied courtesy of Dr. Hugh Lewis, University of Southampton, UK.)

into Earth's atmosphere. Also, peaks in debris density occur in orbital regions where there are lots of spacecraft and where the atmospheric density is too low for *drag sweeping* to be effective in removing objects from orbit. A good example of this are the peaks in debris density at altitudes of around 800 to 1000 km (500 to 620 miles), where there are a large number of Earth observation satellites, and where the atmosphere is too tenuous for drag to be effective in removing the resulting junk caused by operating these spacecraft. Figure 6.10 also gives us an idea of the average spacing between large (greater than about 10 cm in size) objects currently in LEO.

The vertical axis suggests that the peak in debris spatial density is around 2×10^{-8} objects per cubic kilometer. A simple calculation reveals the significance of this obscure statistic. If the objects were distributed evenly, then each one would have its own volume of space in which to wander around, equivalent to a cube about 370 km (230 miles) across. On the one hand, it seems like a huge amount of space for the chunk of debris to get lost in. A cube of this size contains about 50 million cubic kilometers! On the other hand, traveling at typical LEO speed, the object can traverse this space in less than a minute. The bottom line is that space debris is not evenly spaced out in orbit, and debris does come together now and again. However, the low spatial density tells us that this does not happen often; indeed, at the time of this writing, only three collisions between catalogued objects have been verified. The first of these occurred in 1996 between a small French satellite called Cerise and a fragment of an old Ariane launch vehicle.

People who operate shiny, expensive spacecraft in LEO are understandably protective of their investment in orbit, and debris is an obvious threat to their spacecraft's mission. Clearly, if the spacecraft were to be hit by a large chunk of space debris, the impact would be catastrophic, given that relative speeds in orbit are typically on the order of 10 km/sec (6.2 mile/sec). Remember, however, that all of the large objects in orbit are catalogued and their orbits are known. So, using computer simulation, the spacecraft's operators are able to keep an eye on all 8000 or so LEO objects in the catalogue to see if any of them are predicted to make a close approach to their valued asset. This is done routinely in operations rooms around the world. If an uncomfortably close encounter is predicted, the spacecraft's orbit is changed to reduce the threat. This type of maneuver has been performed many times by manned shuttles in orbit, as well as by numerous unmanned spacecraft in LEO.

In addition to these large objects in orbit, there are huge numbers of smaller debris objects in near-earth orbits. At the small end of the size spectrum, it is estimated that there are tens of millions of objects in the 1-mm to 1-cm size range. Many of these result from explosive breakups in orbit, but they also have more benign origins, associated with the degradation of spacecraft surfaces exposed to the space environment. The effects of solar ultraviolet radiation and atomic oxygen erosion, combined with repeated thermal cycling on each orbit (due to the spacecraft being exposed to the heat of direct sunlight followed by extreme cold when in Earth's shadow), causes paints and thermal blankets to become brittle. Over time, flakes of material peel off, leaving a wake of small debris particles around the spacecraft as it orbits the Earth. This does not pose any threat to the spacecraft itself, other than the gradual process of general deterioration. But if these small particles are encountered by spacecraft in other orbits, they can impinge on them at high speeds, typically 10 km/sec (6.2 miles/sec). One well-known consequence of this is the frequent need to replace space shuttle windows, due to damage caused by paint flake cratering.

The space debris that poses perhaps the greatest threat to spacecraft is the intermediate-sized objects in the range of 1 to 10 cm, of which it is estimated that there are a few hundreds of thousands in orbit. This is because they are generally too small for their orbits to be determined (and thus they are not in the catalogue), but they are large enough to deliver a lethal blow to an operational spacecraft. Obviously, if you cannot predict when they are coming, then you cannot perform orbit change maneuvers to avoid potential impacts.

Despite the large number of objects in this size range, there is an awful lot of space in LEO, so thankfully the chances of an impact with this size of object is still very small. For objects smaller than about 2 cm (about 1 inch) in size, however, it is possible to adopt a different protection strategy— *shielding*. This has been used extensively to protect the International Space Station (ISS) from debris impact, and may be used more widely in future unmanned spacecraft in orbital regions where the debris impact threat is considered significant. How can a shield be devised to stop a 2-cm chunk of aluminium traveling at speeds that make a high-velocity bullet look harmless? The idea for such a shield was first proposed by Fred Whipple in 1946, and the simplest form of the *Whipple shield* is shown in Figure 6.11. The construction is straightforward; a bumper shield plate is fixed to the

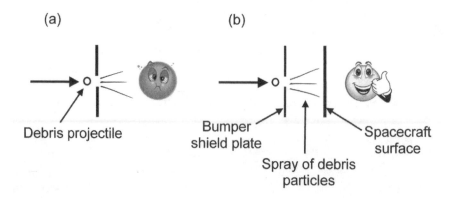

Figure 6.11: Encounter with a debris projectile (a) without a Whipple shield and (b) with a shield.

spacecraft's surface, such that there is a gap between them. The idea is that an impacting debris object is disrupted and vaporized by its encounter with the outer bumper plate, producing a spray of smaller debris particles that impact on the inner spacecraft surface. Since the energy of this spray is spread over a wider area of the spacecraft's surface, the chances of penetration are reduced. The effectiveness of the shield can be improved by introducing multiple bumper shield plates, and playing tunes on the separation distances between them. Generally, however, spacecraft designers would prefer not to implement such shields, as they introduce complexity and mass into the design, with the potential to increase mission costs.

Summary

The space environment is complex, especially the aspects associated with the dominant influence the Sun has over the near-Earth environment—the so-called solar–terrestrial interaction. Several textbooks could be written to describe the decades of research that have been done in this area. In writing this chapter, it has been a struggle to keep it brief, and it was difficult to decide what to leave out. Thus, the chapter focuses on the highlights. However, despite this disclaimer, there is enough here to indicate the importance of environment on space vehicle design. The influence of environment in design is not so apparent in the overall spacecraft configuration, as it is for an automobile or airplane. But it can be seen that much research and engineering trial and error have been needed to develop appropriate manufacturing techniques and materials to give the spacecraft a sporting chance of surviving in orbit long enough to fulfill its mission objectives.

Spacecraft Design

Basic Spacecraft Design Method

IN this and the next couple of chapters, we discuss how spacecraft are designed, and what physical factors influence (or drive) the design of the major elements that make up the vehicle. These major elements are referred to as *subsystems*. The process of spacecraft design is all about how these elements are designed and how they are integrated to produce a total spacecraft system capable of achieving the mission objectives and of surviving the damaging features of the space environment that we cited in the last chapter.

Orbit selection was discussed in Chapter 2, and the logic of the method used there is relevant to our discussion now. To review briefly, the process begins with the definition of the spacecraft's *mission objective,* the formulation of a precise statement defining the prime purpose of the spacecraft. This might be something like "the provision of high-resolution imagery of Earth with global coverage," for example. The next step is to choose the *payload instruments* or equipment required to achieve the objective; in this example it would be the cameras required to produce the images of the ground from orbit. The third step is the development of a *payload operational plan:* How does the payload hardware need to operate to best achieve the objective? In Chapter 2, when we discussed orbits, these requirements included where the payload needed to be physically located to maximize its effectiveness, which led naturally to the selection of an appropriate mission orbit—a near-polar low Earth orbit (LEO) in this case. However, this same logic also leads us to the requirements for the design of the subsystem elements of the spacecraft.

The payload is the most important part of the spacecraft; without it, the objectives of the mission cannot be achieved. The subsystems are there purely to support the payload in its operation. Thus the design of each subsystem is driven by what it needs to do and what resources it needs to provide to ensure that the payload does its job effectively. For example, the payload will need a certain amount of electrical power to operate, and so the

G. Swinerd, *How Spacecraft Fly: Spaceflight Without Formulae,*
DOI: 10.1007/978-0-387-76572-3_7, © Praxis Publishing, Ltd. 2008

design of the electrical power subsystem—the size of the solar panels and batteries on board—is governed by this payload requirement. And this type of logic extends to define the design requirements for all the other subsystems as well.

We have mentioned the spacecraft's subsystems but we have not defined them and explained what they do. All spacecraft are comprised of these basic subsystem elements, and Table 7.1 lists the main ones and the functions they fulfill in supporting the payload in its operation.

Table 7.1: The main spacecraft subsystem elements and their function

Subsystem	Function
Payload	To fulfill the mission objective, using appropriate payload hardware (e.g., camera, telescope, communications equipment—depending upon the objective).
Mission analysis	To select the launch vehicle that will launch the spacecraft, to select the best orbit for the spacecraft to achieve the objectives of its mission, and to determine how the spacecraft will be transferred from launch pad to final orbital destination.
Attitude control	To achieve the spacecraft's pointing mission (e.g., to point a payload telescope at a distant galaxy, to point a solar panel to the Sun to raise electrical power, to point a communications dish at a ground station)
Propulsion	To provide a capability to transfer the spacecraft between orbits, and to control the mission orbit (see Chapter 3) and the spacecraft attitude (see Chapter 8), using on-board rocket systems
Power	To provide a source of electrical power to support payload and subsystem operation
Communications	To provide a communications link with the ground, to downlink payload data and telemetry, and to uplink commands to control the spacecraft
On-board data handling	To provide storage and processing of payload and other data, and to allow the exchange of data between subsystem elements
Thermal control	To provide an appropriate thermal environment on board, to ensure reliable operation of payload and subsystem elements
Structure design	To provide structural support for all payload and subsystem hardware in all predicted environments (especially the harsh launch vehicle environment)

A few comments on Table 7.1: First, some engineers prefer not to classify the payload as a subsystem. They like to divide the spacecraft into two parts, and distinguish the payload from the spacecraft *platform* (or service module), the latter being the part of the vehicle containing all of the supporting subsystems. Second, mission analysis is sometimes not considered to be a subsystem, as there is no piece of hardware on board the vehicle that can be identified with this. However, I have included both payload and mission analysis in the table to reflect the structure of a typical spacecraft project design team (see next section). As we will see, all the areas in Table 7.1 are represented by design engineers in such a team, as they all have a profound influence on the overall design process. Third, *telemetry* is mentioned in the table in the communications subsystem section. This is essentially health-monitoring data, generated on board the spacecraft by sensors distributed around the vehicle. These sensors check the state of the spacecraft's components and issue a warning if problems occur. These data are down-linked as telemetry to the spacecraft operations room and are displayed on the operators' computer screens, so that action can be taken if trouble arises.

We can summarize the process described so far with the block diagram in Figure 7.1. Starting with the mission objective at the top and working down, we can decide what payload instruments we need, and how they are to be operated to achieve the objective. Then, once we have the payload, we can look at the resources it needs from the subsystems to operate successfully. Referring to Figure 7.1, the payload will require a particular amount of electrical power, for example, and this will lead to the design of the power subsystem (obviously, the subsystems will also need electrical power to operate, and so in this case the power subsystem design is not just dictated by the payload interface). If the payload requires pointing, such as a telescope or an Earth-observation imaging camera, then the accuracy and stability of pointing will govern the design of the attitude control subsystem. The payload will also generate data, perhaps in the form of pictures from an imaging payload. These data will either be stored on board or transferred directly to the communications subsystem for down-linking to the ground. The rate at which the payload generates the data, and the overall amount of data, will govern the design of the on-board data handling (OBDH) subsystem, which deals with these processes. The data rate generated by the payload then needs to be down-linked by the communications subsystem over large distances to a receiving ground station, and again this leads to the design requirements for the spacecraft's communications subsystem. Also, certain payloads may have to be maintained within a strict temperature range to work properly, and this will govern how the thermal control subsystem is designed.

The design of the subsystems is also influenced by other factors. In

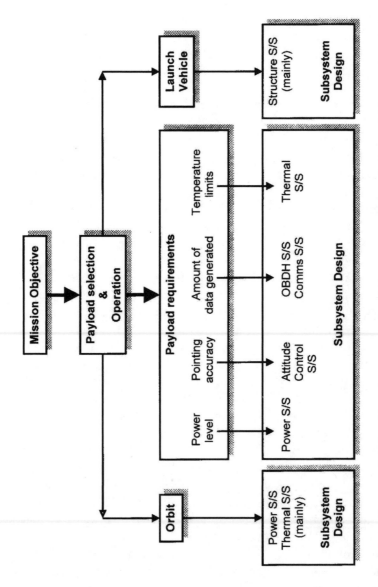

Figure 7.1: A block diagram showing how the spacecraft subsystems are designed. Note: OBDH, on-board data handling; S/S, subsystem.

Chapter 2 we saw how the payload operation led to the choice of mission orbit for the spacecraft, and this is indicated on the left-hand side of Figure 7.1. It is also the case that the orbit itself impacts on the design of the subsystems. For example, once the orbit is specified, the mission analysts can calculate the eclipse period for the spacecraft, which is the time that the spacecraft spends in darkness on each orbit revolution of the Earth. If the satellite is using solar panels to generate electricity during the sunlit part of the orbit, and batteries otherwise, then the design of these components of the power subsystem is greatly influenced by eclipse period of the mission orbit. Similarly, the eclipse period will dictate the amount of direct solar heating (and cooling while in darkness) the spacecraft will encounter on each orbit, which in turn will affect the thermal control subsystem design.

We saw in Chapter 5 how the harsh environment of the launch vehicle was the most influential input to the structure design of the spacecraft, and this is shown on the right-hand side of the Figure 7.1. The design method is not at all mysterious, and is basically applied common sense!

The Spacecraft Design Process

How is this methodology of spacecraft design played out in an industrial setting? The process is very people-intensive, and as such some might observe that it is not quite as objective as you might expect, particularly in the early stages when feasibility and preliminary design issues are addressed. However, we will come back to this perhaps slightly contentious statement later. It is important to set the design method that we have discussed so far in the context of the overall spacecraft development. Spacecraft project activities are traditionally divided into a number of phases, as listed in Table 7.2, taking us from preliminary design through to orbital operations.

Most of what we have said so far falls into phase A, preliminary design, and we do not get much beyond that phase in this book. To get a feeling for how this part of the design is done in a real spacecraft project situation, let's suppose that a company has landed a contract for the phase A study for a particular spacecraft. The process of preliminary spacecraft design that takes place in this phase is sometimes referred to as *spacecraft system engineering*. Definitions of what this means vary, but one possibility is along the lines of "The science of developing an operable spacecraft capable of meeting the mission objectives efficiently, within imposed constraints, such as mass, cost, and schedule." This sounds complicated, but the main job is to design the spacecraft as a collection of subsystems in such a way that, when

Table 7.2 The design and development phases of a spacecraft project

Phase		Duration	Activities
A	Preliminary design and feasibility	6 to 12 months	Creation of a preliminary spacecraft design, and project plan in terms of schedule and cost; the identification of the key technology areas that may threaten feasibility
B	Detailed design	12 to 18 months	Conversion of the preliminary design into a baseline technical solution, including detailed system and subsystem designs; development of a detailed program for subsequent phases
C/D	Development, manufacture, integration, and test	3 to 5 years	Development and manufacture of flight hardware; integration of spacecraft, and extensive ground testing
E	Flight operations	Orbital lifetime	Delivery of spacecraft to launch site; launch campaign; early orbit operations; mission orbit operations; end-of-life disposal from mission orbit

Note: The phase A to D durations are estimates, and vary according to the type of spacecraft.

they are integrated, they produce a total spacecraft design that can efficiently (or even optimally) achieve the objectives of the mission.

Space system engineering is a discipline that differs from spacecraft system engineering. One of the key issues about any kind of system engineering is where you draw the boundary. The focus of this book is concerned with the design of the spacecraft itself (although we do briefly get into launch systems), so we draw a boundary around the spacecraft and focus on it as the system. Space system engineering, on the other hand, draws a wider boundary, not only around the spacecraft but also around all the other parts of the project, such as the ground stations involved in operating the spacecraft and receiving its data. The following discussion does not address these other parts, focusing firmly on the spacecraft.

Returning to the process of spacecraft system engineering, the means of doing this involves forming a design team or committee composed of subsystem engineers, usually with a system engineer as the team leader or chairperson. Each of the spacecraft's subsystems is represented by one or more engineers who are experts in that particular part of the spacecraft. The team leader does not have the same depth of knowledge in any particular subsystem as the team members, but as a system engineer he or she has a breadth of knowledge across the whole system to help in the integration of the overall design.

The traditional method of progressing the design involves lots of off-line analysis and design work by the subsystem engineers, punctuated by numerous meetings of the design team, in which results are discussed and designs are progressed and integrated (it seems odd to be talking about "traditional" methods in spacecraft design, but it is justified by the fact that the space age is half a century old now). This latter aspect is very important. Clearly, each subsystem specialist could work in isolation to produce the most wonderful design solution for his or her own particular corner of the spacecraft. But if it does not integrate with everyone else's subsystem designs, then it is effectively useless.

The team members soon realize that this is design by committee, and that compromise is required by everyone to achieve success at the end of the day, in terms of an integrated spacecraft design. My earlier comment about the objectivity of the process is relevant here. Given how people-intensive it is, spacecraft system engineering could be redefined as the science (*and art*) of developing an operable spacecraft. This work can be a bit of an art form at times, as the outcomes are governed by the dynamics of the team and the interaction among its members. The subsystem specialists have to accept that their own designs will be influenced and modified (perhaps not to their liking!) by inputs from other subsystem or payload specialists.

The other important aspect of this design process is that it is *iterative*. The design team will arrive at an initial design for the spacecraft, but the review of the design will point to areas of the design that can be improved upon considerably, or that may be problematic. The design process is then reviewed—iterated—to overcome these issues in a new design. But the new design may have problems too, and so the process continues until it converges in an acceptable final design.

Over the past decade or so, this traditional method has been transformed by the introduction of computer technology into the process. The basic underlying structure of the design team is still there, but now the team is collocated for the duration of the phase A study in a purpose built design studio equipped with computer work stations. It looks like a miniature version of mission control! Figure 7.2 shows a plan view of such a facility at ESTEC (the European Space Agency's technical head quarters in the Netherlands). Each of the workstations is dedicated to the design of a particular spacecraft subsystem, and as such is loaded with appropriate software to allow operators to do whatever analysis they need to do to

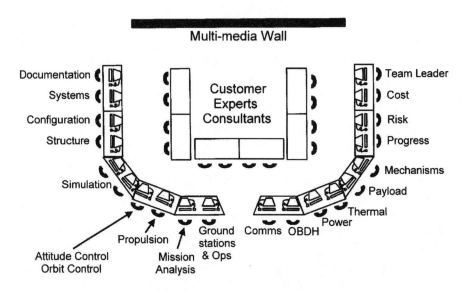

Figure 7.2: A schematic of the computer workstation layout of the European Space Agency's concurrent engineering design facility at ESTEC in the Netherlands. The development of the facility began in 1998. (Backdrop image courtesy of European Space Agency [ESA].)

develop the design. The subsystem engineers comprising the team are now seated at the workstations; for example, the mission analyst sits at the mission analysis workstation, the power engineer at the power workstation, and so on, with the team leader controlling the process from his or her own workstation. The technique is called *concurrent engineering design* (CED), and is being used not just in the space industry but across a broad range of industries involved in the design of complex machines, such as automobiles and airplanes.

At the heart of the process is a central computer database that holds the current design of the spacecraft in memory. Every time team members update the design of their subsystem, the update is relayed to the central database. The details of this change are then available immediately to the other team members, and the impact of the change on the design of the other subsystems can be assessed quickly. The main benefits of this technique are the speeding-up of the process of design iteration that we mentioned above, and an improvement in the process of design integration. The use of CED has typically shortened the time it takes to perform a phase A study of a spacecraft from 6 months or more to 1 or 2 months. However, the introduction of computers into the process has *not* replaced the need for excellent subsystem engineers as design team members. They are still vital in the CED facility to check that the computer output makes sense, and to guide the process to a successful conclusion.

Spacecraft Engineering: The Final Frontier?

Most people, I believe, consider spacecraft engineering to be at the cutting edge of technology, so it comes as something of a surprise that it is often very conservative in its approach. On the one hand, we have subsystem engineers who are always striving to develop new ideas and technologies in their specialist area, to improve the performance of the spacecraft, while at the same time reducing its mass and power requirements. This kind of innovative engineering is the life-blood of the sort of person who becomes a talented and dedicated subsystem specialist. On the other hand, new ideas bring with them questions about their feasibility: Will they work in orbit? How long will it take and how much will it cost to test a new idea to answer these questions to the satisfaction of the project managers? Essentially such innovations introduce risk into the situation, and this kind of risk poses a threat to program schedules, with consequent impact on project costs. Thus we end up in a situation where design solutions that have flown a hundred times before are good, and are automatically adopted in spacecraft designs. This kind of philosophy is common in the design of commercial spacecraft, such as communications satellites. The competition for business among commercial companies in designing and manufacturing these satellites is very stiff, and a particular contractor will invariably adopt conservative engineering solutions in order to minimize cost and schedule in their bid to win the contract. In such situations the time scales for the development phases shown in Table 7.2 are considerably reduced. It is a tough business to be in!

How does the technology progress in spacecraft engineering? It is often the case that innovative engineering is required to build payload instruments for science spacecraft, in order that the boundaries of what has been achieved before can be stretched. However, even on such satellites, the technologies used in the subsystem design—in the spacecraft platform—may be decades old. This is not meant to be a criticism, but may indeed be of benefit in improving the overall reliability of the spacecraft system.

An increasingly popular way of flight-testing new subsystem technologies is through the use of small satellites. There is debate about what is meant by the word "small" in this context, but for our purposes let's think about satellites on the order 50 kg or less. At this sort of mass, and with the continuing trend of miniaturization of computer processors, relatively complex and capable small satellites can be built as flight demonstrators for new technologies. The key to this development is their relatively low cost. In the launch business, satellite mass generally equates to launch costs. Since a small satellite can hitch a ride on a launch vehicle as a minor partner, the cost of launch is hugely decreased. Other factors contributing to their low cost are a short design, build, and test period, and a less complex ground system and operations. Given this low financial risk environment, the consequences of a failure in testing new technologies in orbit are significantly reduced, making the proposition more attractive.

Examples of Current Spacecraft

We conclude this chapter with some examples of current unmanned spacecraft and their applications areas. The main features and subsystem areas (see Table 7.1) of these satellites are shown in figures and described in tables. To put the mass of these spacecraft into perspective, it is helpful to bear in mind that 1000 kg is a metric tonne; a typical automobile weighs about 1¼ to 1½ metric tonnes, and a double-decker bus weighs about 10 metric tonnes. Because the life span of these spacecraft is short, the ones cited here will be relegated to history in a few years. But this book would be incomplete without giving the reader an idea of the mass, size, and appearance of currently flying spacecraft. The particular satellites chosen to represent the different application areas are as follows:

- Communications: Intelsat 8 (Fig. 7.3 and Table 7.3)
- Remote sensing: SPOT 5 (Fig. 7.4 and Table 7.4)
- Science
  - Observatory: Hubble Space Telescope (Fig. 7.5 and Table 7.5)
  - Interplanetary exploration (Cassini/Huygens; Fig. 7.6 and Table 7.6)

Figure 7.3: Artist's impression of the Intelsat 8A communications satellite in geostationary Earth orbit. (Image courtesy of Lockheed Martin.)

Table 7.3: The main features of the Intelsat 8 spacecraft

Description:	Intelsat 8 spacecraft is a typical geostationary Earth orbit (GEO) communication satellite, providing predominantly an intercontinental telephone communication service.
Launch mass:	3250 kg
Dry mass (without fuel):	1540 kg
% of launch mass:	47%
Fuel mass:	1710 kg
% of launch mass:	53%
Approximate size:	Central body is a box 2.2 $\times$ 2.5 $\times$ 3.2 m
Orbit type:	GEO
Height:	35,790 km
Inclination:	0 degrees
Power (beginning of life):	4.8 kW
Payload	
Mass:	460 kg (estimated)
Mass (% of dry mass):	30% (estimated)
Performance:	The communications payload can carry typically 22,000 telephone calls and 3 color TV broadcasts simultaneously

Figure 7.4: Artist's impression of the SPOT 5 Earth observation satellite in a near-polar low Earth orbit. (© CNES/ Artist David Ducros.)

Table 7.4: The main features of the SPOT 5 spacecraft

Description:	The SPOT series of spacecraft was developed as a French national space program. The spacecraft were designed and manufactured by the French space agency CNES (Centre National d'Étude Spatiales).
Launch mass:	2760 kg
Dry mass (without fuel):	2600 kg
% of launch mass:	94%
Fuel mass:	160 kg
% of launch mass:	6%
Approximate size:	2 × 2 × 5.6 m
Orbit type:	Near polar LEO
Height:	822 km
Inclination:	98.7 degrees
Power (beginning of life):	2.5 kW
Payload	
Mass:	1400 kg
Mass (% of dry mass):	54%
Performance:	Provides images of the ground that are 120 km across, with a resolution of 10 m

Figure 7.5: Photograph taken by shuttle astronauts of the Hubble Space Telescope in a near-equatorial low Earth orbit. (Image courtesy of National Aeronautics and Space Administration [NASA].)

Table 7.5: The main features of the Hubble Space Telescope

Description:	Named after Edwin Hubble, who discovered the expansion of the universe in the 1920s, this orbiting telescope has revolutionized all aspects of observational astronomy.
Launch mass:	10,840 kg
Dry mass (without fuel):	10,840 kg
% of launch mass:	100%
Fuel mass:	0 kg
% of launch mass:	0%
Approximate size:	A cylinder 13 m long × 4.3 m diameter
Orbit type:	Low inclination LEO
Height:	Approximately 600 km
Inclination:	28.5 degrees
Power (beginning of life):	5 kW
Payload	
Mass:	1450 kg
Mass (% of dry mass):	13%
Performance:	The main element of the telescope is a mirror 2.4 m in diameter that can see objects just 120 m across on the moon's surface

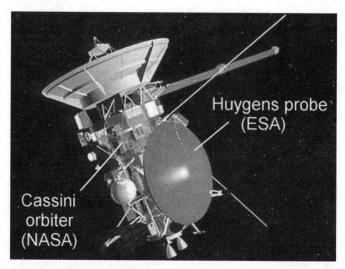

Figure 7.6: The Cassini/Huygens spacecraft configuration. The spacecraft entered orbit around the planet Saturn in July 2004. (Artist's impression by David Seal. Backdrop image courtesy of NASA/Jet Propulsion Laboratory [JPL]—Caltech.)

Table 7.6: The main features of the Cassini/Huygens spacecraft

Description:	The spacecraft is made up of two main parts: Cassini, which is the Saturn orbiter built by NASA, and the Huygens probe built be the European Space Agency (ESA). The Huygens probe successfully landed on Saturn's moon Titan in January 2005.
Launch mass:	5630 kg
Dry mass (without fuel):	2490 kg [2150 kg (orbiter) + 340 kg (probe)]
% of launch mass:	44%
Fuel mass:	3140 kg
% of launch mass:	56%
Approximate size:	6.8 m high, with a 4-m communications dish
Orbit type:	Saturn orbit
Height:	Orbit height continuously modified using swing-bys around Saturn's moons
Inclination:	Near-equatorial
Power (beginning of life):	815 W (using RTGs; see power section in Chapter 9)
Payload	
Mass:	670 kg [330 kg (orbiter payload) + 340 kg (probe)]
Mass (% of orbiter dry mass):	31%
Performance:	The images of Saturn and its moons have been particularly spectacular!

Subsystem Design: I Like Your Attitude

IN the last chapter we discussed the overall process of spacecraft design, and showed where the requirements for the design of the spacecraft subsystems come from (see Table 7.1 and Figure 7.1). The subsystems are there simply to support the payload in achieving the spacecraft's mission objective, so that the resources they need to do this job governs the way they are designed. In this chapter and the next, we take a closer look at the design of the major spacecraft subsystems to identify the main drivers that dictate their design. (The mission analysis subsystem was discussed in Chapters 2, 3, and 4.) Discussing the design of each subsystem will help explain why spacecraft look the way they do.

This chapter discusses the attitude control subsystem (ACS), which is perhaps the most complex of the subsystems and is very influential in determining the overall shape of the spacecraft. *Attitude control* may sound like a sinister governmental plot out of George Orwell's novel *1984*, but in the context of spacecraft engineering it refers to controlling the rotation of the vehicle.

Attitude Control Subsystem

What Does the ACS Do?
Table 7.1 in Chapter 7 stated the purpose of the ACS: "to achieve the spacecraft's pointing mission." What does this mean? Most operational spacecraft in orbit have payloads that require pointing. For example,

- a communication satellite (comsat) needs to point its communications dish(es) at a ground station to receive and transmit the stream of telephone conversations that are using the system,
- an Earth observation spacecraft needs to point its imaging cameras at the targets of interest on the ground, and
- a space observatory needs to respond to ground commands to point a telescope at particular objects (planets or galaxies) in the sky.

G. Swinerd, *How Spacecraft Fly: Spaceflight Without Formulae,*
DOI: 10.1007/978-0-387-76572-3_8, © Praxis Publishing, Ltd. 2008

So, the ACS addresses the pointing, or rotation, of the spacecraft. In earlier chapters we talked a lot about orbits, focusing on how the spacecraft's center moved along its orbit. By contrast, in this section about the ACS, we are not concerned with the motion of the spacecraft's center along its orbit, but rather with the way the spacecraft rotates *about* its center. We can imagine ourselves being in the same orbit as the spacecraft, just a few meters away, and watching it rotate in response to commands to point the payload instruments. The word *attitude* describes the position of the spacecraft in a rotational sense, and a *change in attitude* therefore implies a rotation. What causes the spacecraft to rotate (or to stop rotating)? In the earlier chapters on orbits, we saw that it was *force* that changed the orbit along which the spacecraft moves. In contrast, it is *torque* that causes the spacecraft to rotate about its center.

We discussed torques in Chapter 3, when we looked at gravity anomaly orbit perturbations. A torque is a rotational force such as the one we apply to remove a bolt from a wheel when we change a flat tire. We apply a force of a number of Newtons in a rotational sense by pushing down on the end of the handle of the wrench. The size of the applied torque is based not only on the amount of force exerted but also on the length of the handle of the wrench. The longer the handle, the greater the *moment arm* and the more torque there is. It is easier to loosen the wheel bolts if the handle of the wrench is lengthened. Effectively, this increases the amount of torque, without requiring us to apply more force on the handle. The magnitude of the torque is given by the force times the moment arm, and has units of Nm (Newton meters).

The main job of the ACS is to control the rotational state of the spacecraft by using on-board devices called *control torquers* that produce torques on command to rotate the spacecraft. The most obvious kind of control torquer is a pair of thrusters fired in such a way as to produce a rotation. We will discuss thrusters later, when we talk about the propulsion subsystem, but essentially they are small rocket engines (small enough to be held in the palm of your hand), each producing a thrust force on the order of a few Newtons (a force of 1 Newton is about the weight of a small apple). They are located in clusters around the exterior of the spacecraft, and can be fired in opposed pairs (see Figure 8.4) to produce a torque, and therefore a rotation of the spacecraft. Other kinds of control torquers are discussed later. However, in the interim, it is helpful to briefly describe the main functions of the ACS, as these are the aspects that lead to the ACS design:

- To achieve the pointing mission of the payload, in terms of directions and accuracy. A comsat, for example, may need to point its antenna at a ground station with an accuracy of 0.1 of a degree, whereas a space observatory such as the Hubble Space Telescope may need to point the telescope at a galaxy with an accuracy of less than an *arc second*. If we divide a degree by 60, we get an arc minute, and if we divide an arc minute by 60 we get an arc second. So an arc second is a tiny angle, being 1/3600th of a degree!

- To achieve the pointing requirements of other subsystems, a process sometimes called *housekeeping*. For example:
  - Pointing a solar panel at the Sun to generate electrical power
  - Pointing an antenna at a ground station on Earth's surface to downlink payload data
  - Pointing thermal radiators to the cold of space to allow heat to escape (this will make more sense after we look at the thermal control subsystem)
  - Pointing a rocket engine in the right direction before it is fired, so that the correct change in the orbit is achieved
- Overall, to manage the rotational state of the spacecraft, meaning motion and torques *about* the spacecraft's center.

How Does the ACS Work?

Another aspect that influences the ACS design is how the ACS operates to achieve these functions. Figure 8.1 shows a typical ACS operation, and introduces the main hardware components that comprise the ACS. Starting at the top of the figure, torques act on the spacecraft and cause it to rotate. There are two types of torque. First, there are those we apply deliberately to control the spacecraft's rotation using the on-board ACS hardware. These are produced by the control torquers that we mentioned above. Second, there are *disturbance torques,* which are the rotational equivalent of orbit perturbations. As we saw earlier, the orbit motion of the spacecraft was altered by perturbing forces (see Chapter 3). The spacecraft is also disturbed in terms of its rotation by naturally occurring torques, produced by the spacecraft's interaction with its environment. For example, Figure 8.2 shows how a disturbance torque is generated by a spacecraft's interaction with the atmosphere. As we saw in Chapter 3, atmospheric drag causes a decrease in orbit height, but it can also produce a disturbance torque that causes an unwanted rotation in the spacecraft, which needs to be corrected by the ACS.

The torques act on the spacecraft and cause its attitude to change, that is, cause it to rotate. Moving around Figure 8.1, we see that this rotation is detected and measured by *attitude sensors,* which are hardware components of the ACS, usually optical sensors looking out at *reference objects,* such as the Sun or stars. To picture how these work, imagine being inside a box with windows, which is good description of an airplane. Imagine sitting in a

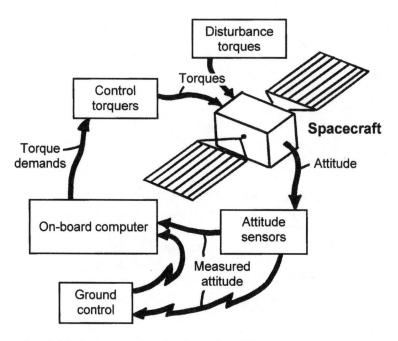

Figure 8.1: A block diagram showing how the ACS operates.

window seat on a night flight. If the cabin lights are down, the stars can be seen easily, and while the airplane is not turning (rotating), the stars look as if they are stationary in the window. However, as soon as the aircraft starts to turn, they appear to move across the window. In the same way, the sensors look out of the spacecraft and interpret movement of reference objects, such as the stars, as a rotation of the vehicle—so allowing measurement of the rotation.

The sensor measurements of the spacecraft's rotation are passed to the on-board computer, which itself can be considered to be another piece of ACS hardware. The measurements are processed by *control software*, which is essentially fancy mathematics coded into the computer to calculate the spacecraft's attitude. This estimate of the attitude is then compared with the attitude required to achieve the pointing mission, and if they differ, the computer's control software calculates what torques are required to correct the spacecraft's attitude. These *torque demands* are then passed to the control torquers (continuing our walk around Fig. 8.1), which then apply the required torques to bring the spacecraft's attitude back to where it should be.

For example, the *pointing mission* for the spacecraft may be to point the antenna of a communications satellite to a ground station to facilitate intercontinental telephone calls. If the satellite is disturbed, the pointing of

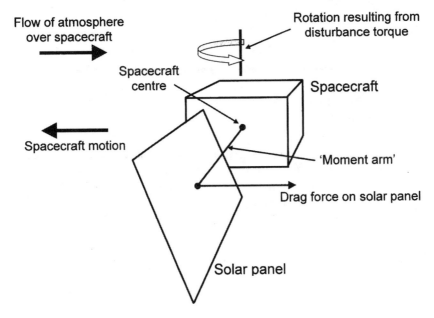

Figure 8.2: An example of a disturbance torque—in this case that produced by aerodynamic drag.

the antenna to the ground station will be affected, but this disturbance will be detected by the ACS sensors. The computer will then process the sensor measurements to command the on-board torquers to correct the pointing error, thus maintaining the communications link. One thing to note about the ACS operation shown in Figure 8.1 is that it operates in a loop, and on most spacecraft it does this automatically many times a second, so that the pointing mission is continuously monitored and maintained. This looping-type operation is referred to as a *feedback loop* by the ACS engineer.

At the bottom of Figure 8.1 there is a ground intervention in this automated process. For example, to operate a space telescope, the astronomers on the ground need to command the telescope to point at a particular galaxy, for example. They can do this by typing the position of the object of interest into a computer on the ground, and this information is then up-linked to the spacecraft and processed by the on-board computer to produce torque demands, which are then executed by the control torquers to rotate the spacecraft to direct the telescope to the required segment of sky.

The main functions of the ACS help other subsystems in their operation (e.g., pointing a solar panel to the Sun to help the power subsystem to do its job). Similarly, if we look at the typical ACS operation outlined in Figure 8.1, we can see that other subsystems help the ACS do its job (e.g., sensors, computers, and control torques need electricity from the power subsystem to work). So, in the design process, the ACS engineer has to work together with many other subsystem engineers, resulting in a very interactive design process.

Attitude Stabilization

One reason why the ACS is considered to be such an important element of the spacecraft is that the type of attitude stabilization used on a particular spacecraft is very influential in determining what it looks like. There are four general types of stabilization, which are illustrated in Figure 8.3. Types 1, 2, and 3 involve spinning all, or a part, of the spacecraft. This spin feature makes the spacecraft's attitude inherently stable; if the spacecraft is affected by a disturbance torque, the change in attitude that results is small. This is a useful feature, as it means the ACS does not have to work so hard to control the spacecraft's attitude.

We can get a good idea of how this inherent stability due to rotation works by looking at a bicycle. The bicycle's tires provide two small points of contact with the ground, each perhaps a couple of centimeters across. The bicycle rider represents a large mass on the top. Other objects with these two characteristics tend to fall over; for example, no matter how hard we try, we cannot balance a nail on its point. But strangely the bike rider is quite happy whizzing along the road without any thought of the bike toppling over. Why? It's basically because the rotation of the wheels give the bike stability; the axes about which the wheels rotate become stiff, in the sense that they want to remain pointing in the same direction. The wheel axles remain horizontal, ensuring that the bike stays upright. As the rider slows down, the wheels' spin rate correspondingly decreases. Eventually the wheels stop rotating, and then the bicycle's stability is lost, and the rider has to put a foot down on the ground to prevent the bike from toppling.

And so it is with a spacecraft. When it spins, the spin axis becomes stiff, tending to make the spacecraft point in a fixed direction. The spin axis becomes less sensitive to disturbances, giving the spacecraft as a whole this characteristic of inherent stability. A spacecraft with this type of spin stability is said to have *momentum bias*. In Figure 8.3, a spacecraft with type 1 stabilization is called a *pure spinner*. These are usually cylindrical in shape, and the whole spacecraft rotates at a rate of a few tens of revolutions per minute (rpm), providing spin stability. The example shown is the Meteosat SG (second generation) satellite, which is a spacecraft that provides satellite pictures for weather forecasts. With this type of stabilization, there is no part of the vehicle or its contents that is not rotating; thus, it is difficult to accommodate payloads that need to point in a fixed direction.

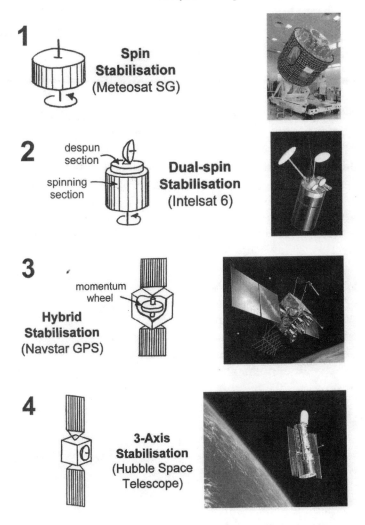

Figure 8.3: The four general types of spacecraft attitude stabilization. All spacecraft fall into one of these categories, some examples of which are illustrated. (Image credits: Meteosat SG image, copyright © ESA; Intelsat 6 image, copyright © Boeing; GPS Navstar 2R image, copyright © Lockheed Martin; Hubble Space Telescope image, copyright © NASA.)

This problem is alleviated by the use of type 2 stabilization. Spacecraft with this stabilization type are referred to as *dual spinners*, and have a cylindrical section spinning at a few tens of rpm (like type 1), giving it spin stability. However, there is a platform mounted on top of the vehicle that is mechanically de-spun, where payloads that need to be pointed in a fixed direction, such as antennas or imaging cameras, can be mounted. The example shown in the figure is the Intelsat 6 communication satellite, which remains (at the time of this writing) the largest example of this type of attitude stabilization.

The third type, which I have called *hybrid stabilization* for want of a suitable label (although this terminology is not used universally by ACS engineers), is quite an interesting arrangement. Here the spacecraft acquires spin stability by mounting a momentum wheel inside the vehicle (see Figure 8.6). In this case, the spin stability is achieved by spinning a small object very rapidly (the wheel rotates at a typical rate of a few thousand rpm), rather than spinning a big object more slowly (the whole or part of the spacecraft rotating at a few tens of rpm, as is the case for the pure and dual-spinners). The mass of the wheel is typically a few kilograms, and its spin rate is maintained at around 6000 rpm. Given that the wheel is rigidly mounted in the spacecraft, its spin stability is transferred to the spacecraft as a whole. Thus the vehicle has the benefit of inherent stability, while at the same time allowing lots of space on the exterior surfaces of the spacecraft to mount payloads and deploy solar panels. The example shown in Figure 8.3 is that of a Navstar GPS satellite, which is part of the U.S. Department of Defense's constellation of satellites used for navigation. Like all such spacecraft using this type of stabilization, the spin stability is not at all obvious to the casual observer, as the mechanism for achieving this (the momentum wheel) is concealed inside the vehicle.

Type 4 is referred to as *three-axis stabilization*. In this case the spacecraft has no significant rotating parts, and thus does not have the inherent stability associated with the other types. As a consequence, the ACS has to work a bit harder to achieve the pointing mission. This lack of stability seems on the face of it to be a disadvantage, but often it is the only suitable option. A good example of this is a space observatory, such as the Hubble Space Telescope shown in Figure 8.3. To achieve its pointing mission, it has to rotate freely to point in various directions, and as a consequence there is no axis in the spacecraft about which it is sensible to employ spin stability. This would make the spin axis stiff, which would make no sense at all if it has to be moved around a lot in the process of pointing the telescope.

The different types of stabilization affect the overall shape or configuration of the spacecraft, which is why the control engineer would say that the ACS is the heart of the spacecraft.

Control Torquers

We have already briefly mentioned control torquers in our walk around the control loop in Figure 8.1. These are items of ACS hardware that are essentially the muscles of the ACS, converting the virtual commands produced by the on-board computer into physical torques that rotate the spacecraft.

Thrusters

Perhaps the most obvious way of producing a control torque is to fire two thrusters in opposite directions, as illustrated in Figure 8.4a. These small rocket engines are set up in groups, called *thruster clusters*, which are located at a number of positions around the spacecraft to ensure that the attitude and orbit control functions can be effectively achieved. By firing thrusters in pairs using different clusters, we can rotate the spacecraft in any direction as shown in Figure 8.4b.

Magnetorquers

When I was a child, I made an electromagnet using a large 6-inch nail, some wire, and a battery. I wound the wire many times around the nail, and

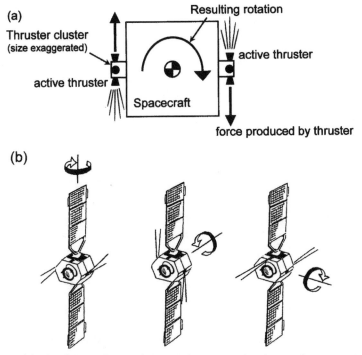

Figure 8.4: (a) The firing of two thrusters in opposed pairs produces a rotational force (torque), causing the spacecraft to rotate. (b) Firing opposed pairs in appropriately chosen thruster clusters allows the spacecraft to be rotated in any direction in three dimensions.

attached the ends of the wire to the battery, and magically when the circuit was made the nail became a magnet. I remember being intrigued by this, and enjoyed the childish pleasure of picking up paper clips and toy cars with this magnet, which I could turn on and off simply by making or breaking the contact with the battery. This simple homemade electromagnet is the basis for another form of control torquer, the *magnetorquer rod*, although the real thing is a little more precisely engineered and a bit bigger! The nail is replaced by a metal rod, usually made from an alloy containing iron, and it can range in length from half a meter to about 2 meters, depending on the size of the spacecraft that needs to be torqued. A considerable length of wire is then wound around this core, giving us an electromagnet that can be activated on command by passing an electrical current through the device.

How is this used to rotate a spacecraft? Consider a compass; the compass needle is simply a magnet mounted on a pivot to allow it freedom to move. It points north because, as a magnet, it tries to align itself with the local magnetic field lines, which at Earth's surface run south to north (see Figure 6.3a in Chapter 6). In the same way, if a current is passed through a magnetorquer, it becomes a magnet, and as a consequence it too will tend to rotate to align itself with the local magnetic field at its orbital position, as shown in Figure 8.5a. Since the magnetorquer rods are firmly attached to the spacecraft (Fig. 8.5b), the vehicle will also share this rotation. So, if we know where we are in orbit, and what the magnetic field is like there, we can produce control torques to rotate the spacecraft on command, by simply passing electrical current through the appropriate magnetorquer. The idea sounds simple enough, but the implementation of this type of control is a fairly complicated business for the ACS engineer. Despite this, however, magnetorquers are commonly used on spacecraft. For example, the Hubble Space Telescope (HST) is equipped with magnetorquer rods to generate torques on the vehicle. An advantage of using magnetorquers on spacecraft like the HST is that they are clean, unlike thrusters that squirt out propellant each time they are used, which could end up contaminating the sensitive telescope optics. Figure 8.5c shows a couple of 1-m magnetorquer rods mounted on a test bed prior to installation in a spacecraft.

Reaction Wheels

Another commonly used type of control torquer is the *reaction wheel*. These are precisely engineered wheels, usually about 15 to 30 cm (6 to 12 inches) in diameter, with a mass on the order of a few kilograms. Their actual dimensions are governed by the size of the spacecraft in which they are installed, and how fast the spacecraft needs to rotate to achieve its pointing mission. To rotate the spacecraft in any direction in three dimensions, three

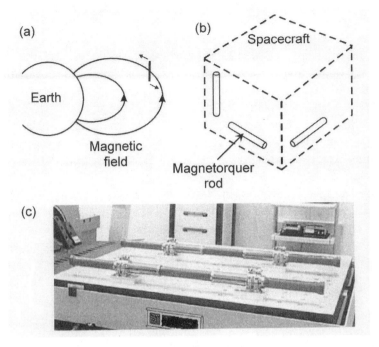

Figure 8.5: (a) When turned on, the magnetorquer rods align themselves with the local magnetic field. (b) The magnetorquer rods are firmly attached to the spacecraft, so that their rotation is shared by the vehicle as a whole. (c) Two magnetorquers under test. (Image courtesy of Dutch Space.)

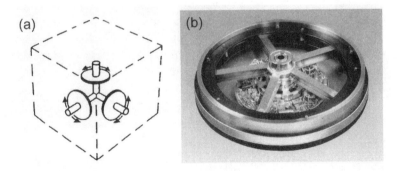

Figure 8.6: (a) Three reaction wheels are mounted rigidly in the spacecraft with their axes perpendicular to one another, to allow rotation of the spacecraft in any direction in three dimensions. (b) A typical reaction wheel. The wheel is usually sealed inside a disk-shaped canister, the lid of which has been removed in the picture. The electronics that control the wheel can be seen underneath the wheel in the base of the canister. (Image courtesy of Rockwell Collins Deutschland.)

wheels are usually employed with their spin axes mounted at right angles to each other, as shown in Figure 8.6a. However, the ACS engineer will usually mount a fourth wheel for redundancy reasons, with its spin axis canted at an angle to the other three, to allow control of the spacecraft in the event of the failure of one of the three primary wheels. An example of what a reaction wheel looks like is shown in Figure 8.6b.

To see how reaction wheels work to rotate the spacecraft, let's focus on just one of them. The wheel is connected to a torque motor, which itself is rigidly attached to the structure of the spacecraft. A torque motor is simply an electric motor that can be used to spin the wheel, a bit like the motor found in a domestic power drill; when we squeeze the drill's trigger, an electric current passes through the torque motor in the drill to rotate the business end—and so it is with the reaction wheel. To rotate the spacecraft around the axis of the wheel, electric current is passed through the wheel's torque motor, and as a consequence the wheel spins. To understand how this causes the spacecraft to rotate, we return to the power drill. If we "tweak" the drill trigger, the chuck and drill bit will spin in one direction, but the handle of the drill kicks back (in a rotational sense) in the opposite direction. This is why astronauts have trouble using power tools when working during space walks; the reaction causes them to rotate as well as the tool, so they have to be firmly attached to the spacecraft to prevent a rather undignified pirouette! It is this same reaction that kicks the spacecraft into rotational motion about the wheel axis, but in the opposite direction to the wheel's rotation. To summarize: To rotate the spacecraft about the wheel's axis, an electric current is applied to the wheel's torque motor. The wheel spins, and as it does so it produces a rotational kick in the opposite direction on the torque motor. Since the torque motor is attached to the spacecraft's structure, this kick is transferred to the spacecraft, which in turn begins to rotate about the wheel axis. This is an application of Newton's third law of motion, which we talked about in Chapter 1: for every action there is an equal and opposite reaction.

But how can the spacecraft be brought to rest again, because once set rotating, the spacecraft will continue to spin forever, as there are no frictional or other forces in space that will stop it. To stop the rotation of the spacecraft, the wheel is brought to rest. The braking (slowing) of the wheel produces a reaction on the spacecraft in the opposite direction that will slow and stop the rotation.

This method of changing a spacecraft's attitude is clean, efficient—elegant even—and just requires a bit of electrical power, which is (usually) freely available through the conversion of sunlight by solar panels. So it doesn't cost propellant, as is the case for thrusters, and generally it has a larger torque capability than magnetorquers. It also works equally well in finely pointing a spacecraft in a particular direction as it does in rotating the vehicle through large angles. Thus, it is the most commonly used technique by the ACS engineer to control the rotation of the spacecraft.

Summary

The choice of the attitude stabilization type has considerable impact on the overall shape and appearance of the spacecraft, and the ACS design is highly interrelated with the design of other subsystems on board the spacecraft. The ACS helps other subsystems in their operation, and conversely the ACS requires their services to allow its own operation.

In terms of hardware, the ACS is composed of sensors, control torquers, and computer processors, but one thing we have not discussed is the complexity of the mathematical control algorithms that are programmed into the onboard computer to make it all work. As a result, the subsystem engineer who works in this area needs to be not only good at the hardware design and its integration, but also a mathematician.

In the next chapter, we discuss the other subsystems and their design.

More Subsystem Design

IT is difficult to say which subsystem on board a spacecraft is the most important, or indeed rank them in order of importance. Subsystem engineers invariably say that their own piece of territory on the spacecraft is the most important. Table 7.1 in Chapter 7 lists the subsystems; a serious failure of any one of them will result in the end of the spacecraft mission. So there is some validity to what subsystem engineers say.

This chapter discusses the remainder of the subsystems, focusing on the main factors that drive their design. As a consequence, I should warn you that this chapter is longer than previous ones. We start with the propulsion subsystem simply because it follows logically from the discussion of orbits in Chapters 2, 3, and 4.

The Propulsion Subsystem

As we recall from Chapter 7, the propulsion subsystem's main functions are to provide a capability to transfer the spacecraft between orbits, and to control the mission orbit (see Chapter 3) and the spacecraft attitude (see Chapter 8), using on-board rocket systems. Although this statement seems elaborate, it is reasonable to say that the propulsion subsystem's job is fairly intuitive; that is, the spacecraft has rocket engines on board that are fired to move the vehicle as appropriate to ensure it reaches its intended destination in near-Earth or interplanetary space. There are effectively two aspects to the propulsion subsystem's job: *orbit transfer,* and *orbit and attitude control.*

Orbit Transfer

Some spacecraft can be placed directly into their mission orbit by the launch vehicle, and thus have no need to perform orbit transfers. However, other spacecraft have to transfer between orbits to reach their final destination. This process of orbit transfer usually involves the use of a large rocket engine onboard the spacecraft, and such a system is referred to as *primary*

G. Swinerd, *How Spacecraft Fly: Spaceflight Without Formulae,*
DOI: 10.1007/978-0-387-76572-3_9, © Praxis Publishing, Ltd. 2008

propulsion. If a spacecraft needs such a system, then the mass of the rocket hardware and of the required propellant has a major effect on the overall mass of the spacecraft. A good example of a spacecraft that requires such a primary propulsion system onboard is one of the communication satellites that we talked about in Chapter 2. They can be injected into a low Earth orbit (LEO) by the launch vehicle, but then need to be boosted to the high geostationary Earth orbit (GEO) before they can begin operations. This is usually achieved by using the *Hohmann transfer.* This strategy, which was invented by Walter Hohmann, is illustrated in Figure 9.1. Hohmann was one of those extraordinary individuals you find in the history of spaceflight. During the early 1900s he was a city architect by profession, in Essen, Germany. But in his free time he devoted his energies to thinking about interplanetary spaceflight. He published his work, including the details of his orbital transfer method, in 1925, at a time when the first Earth satellite was still over a quarter of a century away. His transfer method has been used hundreds of times in placing spacecraft into GEO because it does the job using the minimum amount of rocket fuel, which. means that the overall mass of the spacecraft is minimized and the launch costs can be reduced.

Referring to Figure 9.1, if we assume that the spacecraft is placed initially in LEO by the launcher, how does the Hohmann transfer take us to GEO? To do this, the spacecraft's primary engine is fired twice—once at point 1, to boost it into the geostationary transfer orbit (GTO), and then again at point 2 to push it into the final GEO. The elliptical GTO becomes a sort of bridge, spanning the space between the low orbit and the high mission orbit. Focusing on the first engine firing at point 1, we know from our basic orbits in Chapter 2 that the speed of the spacecraft at this point in the circular LEO

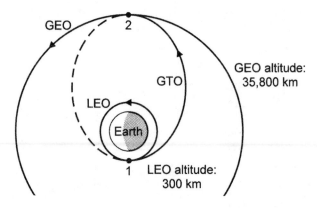

Figure 9.1: The Hohmann transfer between two circular orbits in the same plane. In this case, the spacecraft engine is fired at point 1, to transfer from LEO to GTO. The engine is then fired again at point 2, to take the vehicle into the GEO.

is about 8 km/sec (5 miles/sec). Recalling Newton's cannon on the mountaintop (Chapter 2), we know that the speed at point 1 at the perigee (the low point of an elliptic orbit) of the GTO is higher. This is very much related to our discussion about what happens if we increase the barrel speed of Newton's cannon beyond the circular orbit speed. With some simple calculations we can determine that the speed at point 1 in the GTO is about 10 km/sec (6.2 miles/sec). Thus if the rocket engine firing at point 1 increases the speed of the spacecraft by 2 km/sec (1.2 mile/sec), then the vehicle will transfer from the LEO to the GTO. A similar calculation shows that the speed increase required at point 2 is about 1.5 km/sec (0.9 miles/sec) to transfer between the GTO and the GEO.

The speed change produced by a rocket burn is referred to as a ΔV (pronounced "delta vee"), and one of the prime jobs of the mission analysis team is to calculate the total mission ΔV required to take the spacecraft from launch pad to final mission orbit. Mission analysts spend a lot of time calculating ΔVs because it is directly related to the amount of rocket fuel required. The first person to realize this was Konstantin Tsiolkovsky, another spaceflight visionary who worked as a high school mathematics and science teacher in Russia around the turn of the 20th century. He published his *rocket equation* in 1903, in what was arguably the first mathematical treatise on rocket science. This equation calculates the rocket fuel mass for a particular orbit transfer directly from the corresponding ΔV, which is an important calculation for the spacecraft designer. Mission analysts should also try to minimize the ΔV, because minimum ΔV means minimum fuel mass; in turn, minimum fuel mass means maximum payload mass onboard the spacecraft, which means that the overall effectiveness of the spacecraft in achieving its objective is enhanced. You may recall a similar argument in Chapter 5 when we discussed maximizing the payload of a launch vehicle.

We can use Tsiolkovsky's rocket equation to estimate the rocket fuel mass required for the transfer from LEO to GEO. The total ΔV of about 3.5 km/sec (2.2 miles/sec) requires that about 70% of the initial mass of the spacecraft in LEO needs to be propellant. This leaves only 30% of the initial mass as hardware (payload and subsystems), which poses a problem for the designer. To help overcome this, it is common practice for the launcher to inject the spacecraft directly into the GTO, so that only the 1.5 km/sec ΔV at point 2 is handled by the spacecraft's own primary propulsion. In this case, only about 40% of the initial mass needs to be rocket fuel, giving the designer a much more reasonable 60% of hardware mass to play with.

In our discussion of the Hohmann transfer, we used the LEO to GEO journey as an example. But this type of transfer can be used to move a spacecraft between any two circular orbits that share the same plane, for example, between two LEOs a few hundreds of kilometers apart, where the ΔV might be a few hundred meters per second. The original application that Hohmann had in mind was the transfer between the orbits of two planets, say Earth to Jupiter, where the ΔV might be on the order of 10 km/sec (6.2 miles/sec).

Orbit and Attitude Control

As we saw in Chapter 8, the attitude control subsystem borrows the services of the propulsion subsystem to do the job of attitude control. The small thrusters, fired in opposed pairs (see Figure 8.4), are used as control torquers. These small rocket engines, grouped in clusters, around the spacecraft, are referred to as the spacecraft's *secondary propulsion system*.

The orbit control function, which we discussed in Chapter 3, is also handled by the spacecraft's secondary propulsion. We found that when the spacecraft is launched into the ideal mission orbit, it does not stay there, unfortunately. Perturbing forces due to drag, the Sun's, Moon's and Earth's gravity, and light pressure cause small changes in the orbit, which must be controlled if the mission orbit is to be maintained. To correct the orbit, small changes in the spacecraft's orbital speed are required, and we have seen that this can be achieved by firing small rocket engines onboard the spacecraft. Two small thrusters are fired, not in opposite directions to produce a torque, but this time in the same direction to produce a small ΔV to correct (or control) the orbit, as shown in Figure 9.2.

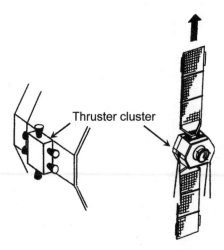

Figure 9.2: Two thrusters are fired in the same direction, to produce a small change in orbital speed to control the orbit against perturbations.

Spacecraft Propulsion Technology

Currently there are two main types of onboard spacecraft propulsion systems: chemical and electrical. Chemical systems, such as the large launcher rocket engines we discussed in Chapter 5, use the chemical energy obtained by burning a propellant/oxidizer mix. This produces a hot and energetic gas, which is then exhausted through a rocket nozzle to produce thrust. Electrical systems, on the other hand, use electrical power to accelerate the propellant gas out of the rocket nozzle. High exhaust speeds can be obtained in this way, but the amount of mass per second that can be accelerated is generally small, unless a large amount of electrical power is used. As a consequence, the thrust of electrical systems is small (usually small fractions of a Newton). This chapter focuses on the more commonly used chemical propulsion systems. Big chemical rocket engines on spacecraft are either solid propellant systems or liquid bi-propellant systems.

Primary Propulsion

Solid Propellant Systems

A typical *solid propellant* main engine on a spacecraft looks like the system shown in Figure 9.3. In concept, it is a simple device, composed of a tank containing solid fuel, an engine nozzle, and some kind of *pyrotechnic device* to light it. Just like the large space shuttle solid propellant booster rockets (it may be helpful to recall the material associated with Figure 5.2 in Chapter 5), once lit, it will burn until the propellant is exhausted, and then it becomes

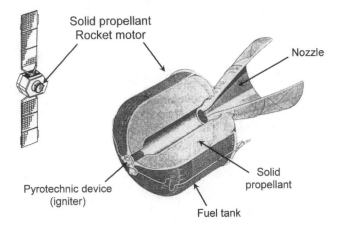

Figure 9.3: Diagram of a typical solid propellant primary engine.

inert and plays no further role in the spacecraft's mission. The engine burn is usually of short duration (a few tens of seconds) and high thrust (tens of thousands of Newtons), producing high accelerations—typically a 0 to 60 mph in 1 second kind of performance! In our orbit transfer example from LEO to GEO in Figure 9.1, the engine burn at point 2 has been performed many times with this kind of rocket engine. To get the right ΔV at this point, the mission analysis team has to do careful calculations, using the rocket equation, to determine how much propellant is needed. This precise amount is then loaded into the motor. Despite the simplicity of the system, one major disadvantage of using a solid propellant motor is the one-shot characteristic; multiple rocket engine burns are not possible.

Liquid Bi-Propellant Systems

Over time, spacecraft missions have become more sophisticated, requiring more than a single primary engine burn. To accommodate this requirement, a *unified liquid bi-propellant system* has become increasingly popular as an alternative to the solid propellant systems. The system is unified in the sense that the spacecraft's primary engine, and the small thrusters that deal with orbit and attitude control are on the same circuit. The big and small rockets onboard the spacecraft share the same fuel supply. The propellant used is in liquid form, and it is a bi-propellant because there are two liquids—a fuel and an oxidizer. The commonly used fuel and oxidizer are monomethylhydrazine and nitrogen tetroxide, respectively. These liquids are *hypergolic,* which basically means that if we mix them together, they spontaneously explode! So to fire the main rocket engine, for example, all we have to do is feed the fuel and oxidizer into the rocket's combustion chamber (see Figure 5.3), and a hot gas is produced explosively, which is then exhausted through the engine nozzle to produce thrust. The same principle is used to operate the small thrusters on board the vehicle. The liquid propellants are held under the pressure of a gas, such as helium, so that the feed system in this case operates by squeezing the liquids down the fuel lines under this gas pressure. Of course, the hypergolic character of the fuel/oxidizer mix means that the plumbing associated with the propulsion system is complex, in order to ensure that the two liquids do not mix before they get to the combustion chambers of the thrusters. The consequences for the spacecraft were this to happen would be catastrophic! Figure 9.4, which shows an example of the plumbing associated with such a system, gives a good idea of this complexity. There is also the issue of the safe handling of the fuel and oxidizer at the launch site, when the spacecraft's fuel tanks are being loaded. These workers have to be equipped with protective, pressurized suits to safeguard them from the effects of an accidental escape of these nasty substances.

Oxidiser tank

Propellant tank

High pressure
gas tank

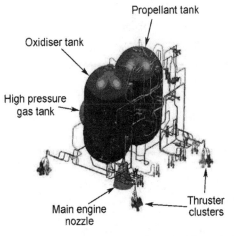

Main engine
nozzle

Thruster
clusters

Figure 9.4: A diagram of the unified liquid bi-propellant propulsion system used on the European Space Agency (ESA) Venus Express spacecraft. As well as the main elements that are labeled, there is also an array of valves, sensors, and regulators to manage the hypergolic fuel safely. (Image credits: left—courtesy of EADS Astrium; right—courtesy of ESA.)

The main engine used in unified liquid bi-propellant systems usually has a much lower thrust than solid propellant primary engines. A thrust level of around 400 Newtons is common, giving a 0 to 60 mph in 2 minutes type of performance for an average-sized spacecraft! Consequently, the duration of an engine firing is often much longer, up to 1½ hours, to give enough time for the required ΔV to be achieved.

Secondary Propulsion

We have already discussed the operation of small thrusters on spacecraft equipped with unified liquid bi-propellant systems. However, spacecraft equipped with solid propellant primary engines, have a need for an independent thruster arrangement, and this is usually a *mono-propellant hydrazine* system. "Mono" implies that this involves a single liquid fuel, as opposed to the fuel/oxidizer combination in the bi-propellant system; this single fuel is fed under gas pressure to the thrusters. To fire a particular thruster, the fuel inlet valve is opened, and the mono-propellant hydrazine is squirted through a bed of chemicals in the thruster, causing an exothermic (heat producing) chemical reaction, which breaks down the liquid fuel into hydrogen, nitrogen, and ammonia gases. These hot gases are then exited through the thruster nozzle to produce thrust. Figure 9.5 shows a typical mono-propellant hydrazine thruster, which fits comfortably in the palm of a hand.

Figure 9.5: An example of a mono-propellant hydrazine thruster, with a thrust level of 5 Newtons. Left: The thruster showing the classic thruster nozzle. Right: The same thruster in its normal operational configuration with collar attached. (Image courtesy of EADS Astrium.)

The Electrical Power Subsystem

As we recall from Table 7.1 in Chapter 7, the main function of the power subsystem is to provide a source of electrical power to support payload and subsystem operation. This job is critical to the overall health of the spacecraft. Just about every type of spacecraft payload and all the subsystems, with the exception of the structure and possibly the thermal control, need a reliable source of electrical power to operate. A failure, or brief interruption, of the power subsystem function can have catastrophic consequences for the spacecraft mission.

In the same way that the spacecraft has primary and secondary propulsion (see the previous section), the spacecraft also has primary and secondary power systems. The *primary power system* is the main source of electrical energy; for example, for Earth-orbiting satellites this is often the conversion of sunlight into electricity using a solar panel (or array). As you have read through the previous chapters in this book, you've seen numerous pictures of spacecraft, and the majority of them are equipped with solar arrays. The *secondary power system* comprises electrical storage devices. In the vast majority of spacecraft, this implies the use of battery technology. However, there are other possibilities, although they are rarely used. For example, a fly wheel can be installed as an alternative electrical storage device. While the spacecraft is in sunlight, solar panel power can be supplied to a torque motor to spin a large wheel. When the spacecraft is in darkness, and the primary power source no longer works, the rotational energy in the wheel can be extracted and converted back into electricity.

A spacecraft is very much like an automobile inasmuch as there is a primary power source combined with a secondary power system. In a car the primary system is the generator that provides electrical power all the time that the engine is turning. The secondary power system is the battery, which allows the car's systems to operate even when the engine is not running. Despite the terminology, the secondary system is very important, in that it provides a means of starting the car and bringing the primary power source on line!

The most common mode of operation for the supply of electrical power on board Earth-orbiting spacecraft involves a solar array/battery combination. While the spacecraft is on the sunny side of the Earth, the solar array operates to produce power for the vehicle's payload and subsystems. At the same time, an extra amount of power is produced by the solar array to charge the battery system. Then, when the spacecraft enters the Earth's shadow on each orbit revolution, the solar array ceases to function and the vehicle's systems are powered by discharging the stored energy in the batteries.

Commonly Used Primary Power Sources

In terms of primary power sources, there are a number of options (including solar arrays), and the main candidates are listed in Table 9.1. Figure 9.6 shows the most suitable primary power source for a spacecraft, given the duration of its mission and its electrical power requirement. From this we can see that, of the six sources listed in Table 9.1, only four are suitable for long-duration space missions, and of these only two—solar arrays and radioisotope thermal generators (RTG)—are commonly used.

Solar Arrays

Recall that in Chapter 6 we (figuratively) went into the garden and presented a square meter of area to the Sun, and found that the Sun's power falling this was roughly 1.4 kW, neglecting the losses that occur due to passage through the atmosphere. In Earth orbit, this *power flux* is essentially a free resource that is just too good to miss, so spacecraft are usually equipped with solar arrays designed to convert some of this solar power into electricity. Most

Table 9.1: Primary electrical power sources used on spacecraft

Type	Usage and principle of operation
Primary batteries	Primary (unrechargeable) batteries are used for short duration missions, for example, to power a launch vehicle during the few minutes' climb to orbit.
Fuel cells	Fuel cells are essentially chemical engines that produce electrical power, with water as a by-product. This makes them particularly suitable for manned missions. The duration of their operation is limited, however, by the requirement to fuel the chemical reaction continuously with oxygen and hydrogen. Fuel cells provide the primary power source for the space shuttle.
Solar arrays	Solar arrays operate by converting sunlight into electrical power. With the intensity of sunlight at Earth (about 1400 W per square meter), solar arrays on Earth-orbiting spacecraft provide about 100 W of useful electrical power for every square meter of solar array surface area. So, although solar arrays are commonly used, they are inefficient (see text for more detail).
Solar dynamic devices	A solar concentrator, such as a large parabolic mirror, is used to focus the power of the Sun to heat a working fluid, such as water. The high-pressure steam produced can then be used to drive a turbine generator—the dynamic part—to produce electrical power. Solar dynamic devices are more efficient than solar arrays, but are generally much heavier. As such, they are only considered for use on large spacecraft, such as space stations. Overall, they are rarely used.
Radioisotope thermal generators (RTGs)	The heat energy obtained from a radioactive material (such as an isotope of plutonium) is converted into electrical energy. RTGs are widely used on spacecraft that operate a long way away from the Sun, where sunlight is so feeble as to make the use of solar arrays impractical. Each RTG is cylindrical in shape, typically 1 m in length and 30 cm in diameter (see Figure 9.8). Such a device is around 40 kg in mass, and produces about 200 W of useful electrical power. A quick calculation (200 W divided by 40 kg) gives a rough estimate of the amount of power produced per kilogram—approximately 5 W. So if large amounts of power are required for payload and subsystem operation, then the mass of the power supply can be significant. There is also political opposition to the

use of RTGs by the "Green" lobby, which fears the effects of the dispersal of radioactive material in the event of a launch failure.

Nuclear reactors   These are scaled-down versions of the nuclear reactor systems found in terrestrial power stations. They are used only for applications requiring large amounts of power—on the order of hundreds to thousands of kilowatts. To date, they have been used rarely, and then only on military space systems such as large active radars for surveillance.

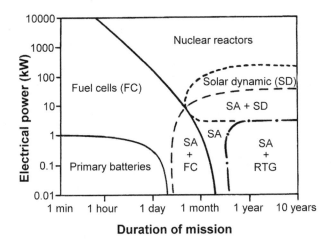

Figure 9.6: The most suitable primary power source is shown, depending on the level of electrical power required by the spacecraft and the duration of its mission. Note: SA, solar array; RTG, radioisotope thermal generator.

solar arrays used for space applications are made out of the semiconductor material silicon, and they have an efficiency of around 10%. This means that of the 1400 W of solar power falling on each square meter of array, only one tenth of this (140 W) is produced as useful electrical power to run the spacecraft's systems.

Unfortunately, there are a number of other factors that have to be taken into account, that further reduce the solar array's efficiency. For example, for best performance the array surface needs to be presented to the Sun so that the Sun's rays fall at right angles to its surface. This requires the array to be accurately pointed by the spacecraft's attitude control subsystem (ACS) (see Chapter 8), and this may be done only to a certain tolerance—say, to within 5 degrees of the ideal. Such a *pointing error* will reduce the array's efficiency.

Another factor is temperature. Arrays get hot while in the Sun, typically up to around 50°C (in Earth orbit), and the hotter they get the less efficient they are; for example, silicon arrays can lose 10% of their electrical performance with a 25°C increase in temperature. A third issue is that the spacecraft's solar array panels cannot be covered completely in useful silicon material. Each panel is usually composed of smaller silicon cells, and the required electrical interconnection between these means that about 10% of the panel area is dead with respect to power generation. The final factor is damage to the array caused by particle radiation (see Chapter 6). After 10 years of operation in GEO, the electrical output from a solar array can be reduced by around 30% of its beginning of life performance.

All these factors vary according to the spacecraft's mission and orbit, but in Earth orbit a useful rule of thumb is to expect about 100 W of useful electricity—enough to run an old-fashioned domestic light bulb—for every square meter of silicon solar array on the spacecraft. Many modern spacecraft, particularly communication satellites, can have power requirements up to 10 kW, which means that an lot of solar array area is needed. Accommodating such large arrays can have a significant impact on the overall configuration of the spacecraft.

The above discussion applies to Earth-orbiting spacecraft, which are effectively at a distance from the Sun of 1 astronomical unit (AU). As we have seen, the Sun's intensity at 1 AU is about 1.4 kW per square meter, but this intensity varies with distance from the Sun. In fact it decreases in proportion to the inverse square of the distance; Chapter 1 discussed the inverse square law in the context of Newton's law of gravity. Basically, the idea is that at twice the distance from the Sun at 2 AU, the Sun's intensity will fall to a quarter ($1/2^2$) of what it was at Earth. At 3 AU it will decrease to a ninth ($1/3^2$), and so on. So, for example, if our mission is to take us to Saturn, which is about 10 AU from the Sun, the solar power flux there is about 1400 W divided by 100 (10^2)—about 14 W per square meter. On top of this, there are all the various inefficiencies of the solar array in converting this power flux into electricity, so we can see that at this kind of distance the use of solar arrays to generate electrical power becomes impracticable.

What is the maximum distance from the Sun where solar arrays can still be used effectively to generate spacecraft power? This is difficult to answer, but I would suggest 5 AU (the distance of Jupiter from the Sun) as the outer boundary. A quick calculation tells us that at 5 AU the solar intensity is 1400 W divided by 25 (5^2), which is about 56 W per square meter. Taking account of the efficiency of the array, we might expect about 6 W of electrical power per square meter of array, which still sounds marginal. But one saving grace here is the temperature of the array, which out in the cold reaches of Jupiter's orbit will typically be less than $-100°C$. As mentioned above, solar arrays become less efficient as their temperature rises, but then by the same token they become more efficient as their temperature drops. Consequently, the array can achieve a useful level of power generation even at these distances. Interestingly, at the time of this writing (2007), the European Space Agency (ESA) spacecraft Rosetta is on its way to rendezvous with a comet in 2014, and the interception will take place at a distance of about 5¼ AU from the Sun where the solar intensity is about 50 W per square meter. The Europeans are generally averse to using RTGs, on environmental grounds, and so the spacecraft is equipped with solar arrays for power generation. However, to generate the required 395 W of electricity to run the spacecraft—equivalent to four light bulbs—64 square meters of solar array are needed! A quick calculation, 395 W divided by 64 m^2, gives about 6.2 W of electrical power per square meter of array. The overall array efficiency (6.2 W/m^2 of electricity divided by 50 W/m^2 of solar intensity) is about 12%, which is just about feasible. It is at these great distances from the Sun where it is advantageous to think about the use of RTGs.

Radioisotope Thermal Generators (RTGs)

To overcome the problems of power generation at large distances from the Sun, the use of RTGs provides a solution. These have been used on many spacecraft (for example, Pioneer, Voyager, Ulysses, Galileo, and Cassini), which have ventured to distant parts of the solar system, at and beyond Jupiter's orbit. The idea is that they take their own energy source with them, in the form of a radioactive material. The radioactivity heats the material, and this heat is converted to electricity through the *thermoelectric effect*. This was discovered, it is said accidentally, by Thomas Seebeck in 1821, and you may have done experiments in school science lessons to demonstrate the effect, using a *thermocouple*. Figure 9.7 illustrates a simple thermo-couple: two wires made of dissimilar metals A and B are formed into a circuit, with some kind of meter to measure electrical current. If junction 1, where the two metals are joined, is heated, and junction 2 is cooled, an electrical current is generated in the circuit. The heat differential produces electricity. The RTG is a more sophisticated device, but this simple thought experiment demonstrates the underlying principle of its operation. Essentially, the temperature difference between the hot core of the radioactive material and the cold of space is exploited to produce electrical power for the spacecraft. The typical size, mass, and power output of a RTG are given in Table 9.1, and Figure 9.8 shows what they look like.

There are disadvantages to using RTGs, some of which we have already mentioned. Given that a typical RTG contains a radioactive material such as

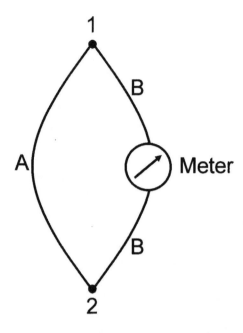

Figure 9.7: A simple thermocouple circuit composed of wires of different metals A and B, and a meter to measure electrical current. When junction 1 is heated and junction 2 cooled, an electrical current will flow in the circuit.

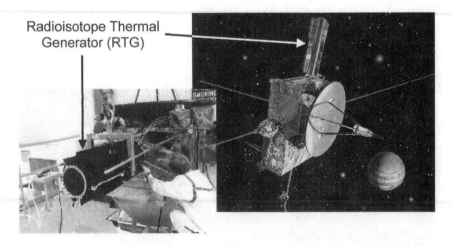

Figure 9.8: The RTG on the ESA/National Aeronautics and Space Administration (NASA) spacecraft Ulysses, shown during spacecraft assembly (left) and in flight (right). (Image credits: left—NASA/Jet Propulsion Laboratory (JPL)—Caltech; right—ESA.)

plutonium, there are safety concerns related to the radiation hazard RTGs pose when the spacecraft is being assembled. This radiation is also potentially damaging to onboard electronics, and so RTGs need to be positioned away from radiation-sensitive equipment. Also, the heat source needs to be intense to produce the necessary electrical performance. As a consequence, RTGs become hot and produce about 10 times more heat power than electrical power. For a typical device, each RTG installed on the spacecraft radiates about 2 kW of heat, which can create problems with the spacecraft's thermal control and during the launch when the spacecraft is confined in a limited volume inside the launcher fairing.

Typical Power System Operation

Despite the diversity of possible electrical power sources (see Table 9.1 and Figure 9.6), the vast majority of spacecraft operate on a solar array/battery combination. This is particularly true for Earth-orbiting spacecraft, which we shall focus on here. Figure 9.9 is a simplified diagram showing how this typical arrangement works. The main feature is the connection of the spacecraft's electrical loads across the solar array (via points 6 and 7), so that the loads are supplied by the array while the spacecraft is in sunlight. However, there is a complication. We know from the above discussion about solar arrays that if a specified area of array is presented to the Sun, it will produce a specified amount of electrical power. But we also know that the spacecraft loads vary; for example, payload instruments will be switched on and off at various times, and subsystem elements such as reaction wheels, or the communications equipment, will run when they are required. So, on the one hand, the solar array output is constant, but then on the other, the

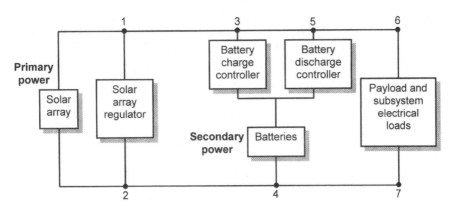

Figure 9.9: A simplified block diagram showing a typical electrical power distribution arrangement for a spacecraft with a solar array/battery power system.

electrical loads it needs to supply are continuously varying. To address this mismatch, a *solar array regulator* is fixed across the solar array (at points 1 and 2). This can be a simple device that takes the excess power not required by the spacecraft loads and dissipates it externally as heat—a bit like an electric bar fire. At the other end of the scale, it can be a sophisticated device, controlled by the onboard computer, which switches patches of solar array area in and out to ensure that the array output matches the loads at any particular time.

While all this is going on, the battery system is also connected across the solar array (via points 3 and 4), so that it can be charged up while the spacecraft is in sunlight; it becomes just another element of the spacecraft's electrical loads. Then, when the spacecraft enters Earth's shadow on each orbit, the electricity for the payload and subsystems can be supplied by the stored energy in the batteries (through points 4 and 5). This process of charging and discharging the batteries needs to be carefully controlled, to ensure that the batteries last long enough to do their job throughout the entire spacecraft mission. The main factors that govern the lifetime of the batteries are the total number of charge/discharge operations needed during the spacecraft's mission, and the amount of stored energy that is taken out of the battery on each such operation. For a typical Earth-orbiting spacecraft, this charge/discharge operation usually occurs once per orbit, and so for a spacecraft in a LEO there are typically about 5000 charge/discharge cycles per year. The *charge/discharge controllers* (points 3 and 5) are essential components to make sure the batteries do not die prematurely. If you think again about the battery in a car, the battery charge and discharge process is not usually controlled in any way; in fact the driver takes this role. Unlike the battery system in a spacecraft, which can last for 10 or 15 years as a consequence of careful control, the battery in a car can expire at the most inopportune moment before its mission (the car's lifetime) is completed!

This section has discussed how the power subsystem achieves the vital role of keeping all the electrical spacecraft systems alive. We now turn to the basics behind spacecraft communications.

The Communications Subsystem

As we recall from Chapter 7, the prime functions of the communications subsystem are to provide a communications link with the ground, to downlink payload data and telemetry, and to uplink commands to control the spacecraft. This is a fairly intuitive task involving the requirement for two-way communication between the spacecraft and the ground. Any activity in orbit, such as taking measurements of the space environment or imaging a galaxy, requires a means of communicating the data to the ground, otherwise there's no point to doing it!

Communications Frequencies

Information on a satellite communications link is carried by electromagnetic (EM) waves; Figure 6.2 in Chapter 6 illustrated the various parts of the EM spectrum. As a consequence, the speed of communication is the speed of light, which is around 300,000 km per sec (186,000 miles per second), so that communication with spacecraft in LEO is effectively instantaneous. However, for a communication satellite in GEO, the altitude of the satellite is around 38,000 km, so that EM waves take just over a tenth of a second to travel from the ground to the spacecraft. This may not seem a lot, but bear in mind that for me (in the United Kingdom) to hold a telephone conversation with someone in the United States requires four such trips for the EM waves; my voice needs to travel up to the satellite, and then down to a ground station in the U.S. My friend's response then needs to make the same return trip, requiring about half a second of travel time. If I talk with someone in Australia, the communications route may involve more than one satellite, so the travel time can be large enough to produce awkward pauses that can disrupt the flow of a conversation. Of course, for interplanetary spacecraft the distances are such that a two-way conversation is not an option; for example, the travel time one-way for EM waves to a spacecraft at Saturn is at least an hour and a quarter.

The wavelength of EM radiation used for satellite communications is between 2 and 30 cm, which is referred to as the microwave part of the EM spectrum (see Figure 6.2). This is also the part of the spectrum used by microwave ovens to heat up dinner in the evening. This type of cooker heats food by bombarding it with EM radiation with a wavelength of typically about 12 cm. Fortunately, the microwave beam from a large satellite communications dish antenna at a ground station is well focused along the axis of the dish and is usually pointed skywards, so it is not a health hazard. Figure 9.10 reviews how the wavelength of EM radiation is defined, but it also shows some other important features. The intensity of the radiation is governed by the *amplitude* of the wave—or the wave height. In terms of the visible part of the spectrum, for example, a bright light has a larger amplitude than a dim one. The figure also shows how the phase of an EM wave is defined. This is measured in degrees, and indicates where you are on the wave along its wavelength; for example, the leading edge of the wave is defined to be at 0-degree phase, the first crest is at 90 degrees, the trough at 270 degrees, and so on. We will see later why this feature of the wave is important.

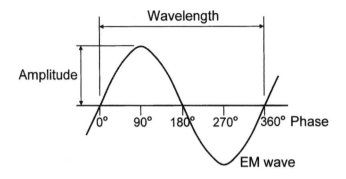

Figure 9.10: The wavelength, amplitude, and phase of an EM wave (see text).

Another aspect of communications is that it is common to talk about the part of the spectrum in terms of *frequency*, rather than wavelength. For every wavelength of EM radiation, there is a corresponding frequency. Generally speaking, short wavelength radiation has a high frequency and long wavelength radiation has a low frequency. This focus on frequency can be seen by looking at a domestic FM stereo radio. A radio station may be listed on the tuning dial as having a frequency of, say, 100 MHz, where the "M" is an abbreviation for "Mega," meaning a million, so we have a frequency of 100 million Hz. "Hz" is short for Hertz, which means cycles per second, named in honor of Heinrich Hertz, a German physicist who made important contributions to electromagnetism. Thus the frequency of the radio station is 100 million cycles per second but maybe this still doesn't mean much to you. Another way of thinking about this is that the wavelength of this signal is such that 100 million wavelengths pass the radio every second traveling at the speed of light. Given that the speed of EM waves is constant, for this to happen a simple calculation shows that the wavelength of a 100-MHz signal must be 3 m.

Satellite communications occur at a higher frequency, and so have a shorter wavelength. Typically the frequencies used are between 1 and 15 GHz, where the "G" stands for "Giga," meaning 1000 million. So the frequencies used are between 1000 million and 15,000 million cycles per second, which (as mentioned above) correspond to a wavelength range of 30 cm to 2 cm, respectively. Why are these particular frequencies used? This choice is dictated by the physics of the atmosphere. For a ground station to talk to a spacecraft, the EM waves must necessarily pass through Earth's atmosphere. However, for frequencies less than about 1 GHz, the energy of the radiation is sapped by charged particles, such as electrons, in the ionosphere. The ionosphere is a region of Earth's atmosphere at heights greater than about 80 km where the atmosphere's molecules of oxygen, nitrogen, etc. are stripped of their electrons by the Sun's ultraviolet radiation. For frequencies higher than 15 GHz, the radiation is absorbed by molecules of water vapor and oxygen in the lower part of the atmosphere. Thus the 1- to 15-GHz frequency range provides a convenient "window" through which the ground station can talk to the spacecraft, and vice versa.

Digital Communications

Another feature of satellite communications is that they are predominantly digital. In terms of satellite communications, *digital* means that effectively all the information in the communications link is converted into a string zeros and ones, each "0" and "1" being referred to as a bit (binary digit) of information. This *binary digital language* is the same as that used by your desktop computer to perform its routine internal operations and to communicate with other computers across worldwide digital networks. In recent years, the use of this digital technique has increased, being applied to television, radio, photography, music, and more. In these areas, and in space communications, the advantage of digital methods is that they produce a better quality of sound or picture, as it is generally easier to distinguish a digital signal from the various sources of interference which compete with it. It is amazing to think how such huge communications and consumer electronics industries can be built on the use of a language that is fundamentally made up of just zeros and ones!

Satellite Telephone Communications

To describe how satellite communications work, let's look in more detail at the process of making an intercontinental telephone call using a GEO communication satellite system. Just as it was in the 1870s when Alexander Graham Bell first invented it, the telephone receiver is an *analogue* device. It operates using continuously varying physical quantities, such as electrical current, without a single digital bit 0 or 1 to be seen anywhere.

In fact, making a telephone call many years ago was a completely analogue process. In this process, the voice produces pressure waves in the air—sound waves—that impinge on a small circular metal disk in the telephone mouthpiece. These waves then cause the disk (sometimes called a diaphragm) to vibrate in sympathy with the voice. Attached to the diaphragm is a lightweight coil of wire, which is positioned adjacent to a permanent magnet. As the wire coil moves up and down with the diaphragm in the magnetic field, a current is induced in the wire (see Chapter 6 for a brief explanation of electromagnetic induction), and this current can be thought of as an electrical representation of the information contained in the voice, that is, speech. This electrical version of the voice then propagates

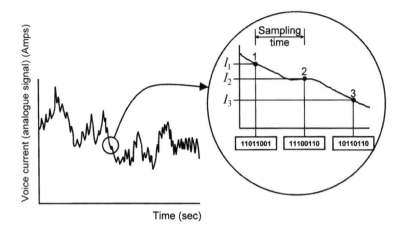

Figure 9.11: A rapidly varying and complex analogue signal is represented on the left. A small section of this is blown up on the right, with the time axis stretched, showing the process of conversion of the analogue signal to a digital one.

down the cable to the destination telephone handset, where the voice current is passed through the diaphragm wire coil of its earpiece. The combination of the voice current in the coil and a magnet in the earpiece causes the diaphragm to move in sympathy with the current. This movement in turn produces sound waves in the air, re-creating an understandable representation of the voice for the recipient of the telephone call.

Because satellite communication is a digital process, at some point the analogue signal—in this case the voice current—needs to be converted into a sequence of zeros and ones to produce a digital signal. This process of converting an analogue signal into a digital one is referred to as *digital encoding* (the method of encoding described below is not currently the most commonly used technique, but it is probably one of the easiest to understand). How can we convert a rapidly varying voice current into a string of 0 and 1 values suitable for transmission to a satellite? The left-hand side of Figure 9.11 is a representation of the analogue signal produced by a telephone, that is, the variation of electrical voice current produced by the telephone mouthpiece. In general, this is a complex and rapidly varying signal, and at any precise moment in time it will have a particular numerical value. The first step, then, is to convert decimal numbers, representing the value of the current at a particular time, into binary numbers, which are simply a sequence of zeros and ones. Table 9.2 shows how the first eight numbers, including zero, can be written in binary digital language. To do so, we need a string of three bits (zeros or ones). The number of zeros and ones in the string required to represent a particular decimal number can be

Table 9.2: The first eight decimal numbers (including zero) represented in binary as strings of three zeros or ones

Decimal number	Binary number		
	2^0 1	2^1 2	2^2 4
0	0	0	0
1	1	0	0
2	0	1	0
3	1	1	0
4	0	0	1
5	1	0	1
6	0	1	1
7	1	1	1

calculated by thinking about powers of 2. Since 8 is 2^3 ($2 \times 2 \times 2$), we require a string of three 0 and 1 values. Continuing in this way, the first 16 (2^4) decimal numbers require a string of four bits, the first 32 (2^5) require a string of five bits, the first 64 (2^6) require six bits, and so on. This is great for people like myself who have quite a few years behind them—instead of having to put fifty something candles on your birthday cake, you only need six candles to write your age in binary. Of course, you do need candles of two colours to distinguish the "0" and the "1". With these kind of jokes on offer you can see that, in my household, birthdays are a whole lot of fun!

As shown in Table 9.2, in binary digital language each bit represents a power of 2, so under the heading "Binary number" we have powers of 2 (2^0, 2^1, 2^2), which take the values of 1, 2, and 4, respectively. We do not often come across a number to the power 0, but any number to the power 0 (e.g., 2^0) takes the value 1. So for each decimal number, we put a 1 where the power of 2 contributes and a 0 where it does not. Here are a couple of examples: on Table 9.2, the decimal number 5 can be written in binary digital language as 1 0 1, because 5 in simple arithmetic is $(1 \times 1) + (0 \times 2) + (1 \times 4)$. Similarly, 6 is 0 1 1, because 6 is $(0 \times 1) + (1 \times 2) + (1 \times 4)$. Thus decimal numbers can be represented as strings of zeros and ones. This strange binary language is what a desktop computer is using all the time to perform its mathematical routines.

In Figure 9.11, the analogue signal—the voice current—is shown on the left-hand side, and we need to convert this into a long sequence of zeros and ones suitable for transmission to the satellite. To show how this is done, we have magnified a small part of the analogue signal on the right-hand side of the figure. At a particular moment, at point 1, the voice current takes a particular value I_1, and this value can be converted into binary digital language. It is common to convert it into a string of eight bits, which means that the voice current value at this point can be assigned to any one of 256 (2^8) possible levels. Then a small fraction of a second later, called the *sampling time,* the value I_2 of the voice current is measured again, and it too is converted to a string of 8 bits. This process continues throughout the telephone conversation, converting the voice current into a long string of 0 and 1 values, as shown in the boxes beneath each point. It is usual to sample the voice current about 8000 times a second (so that the sampling time is 0.000125 second). This short sampling time is needed so that all of the complex and rapidly varying detail of the original voice current is captured in the digital signal. Using this method, the data rate of one digital telephone voice is around 8 × 8000 or 64,000 bits per second—8 bits produced 8000 times each second. This is referred to as 64 kbps, where "k" stands for kilo, meaning a thousand. Computer-savvy readers will be familiar with data rates; for example, a broadband Internet connection might be, say, 5 Mbps, where the "M" stands for Mega, meaning a million.

There is still another step in the process of transmitting a voice to the satellite. The digital bit stream representing a voice now needs to be somehow put onto an EM wave so that it can be transmitted from the ground station antenna to the spacecraft. This step in the process is referred to as *modulation* (Fig. 9.12). Modulation entails putting the information

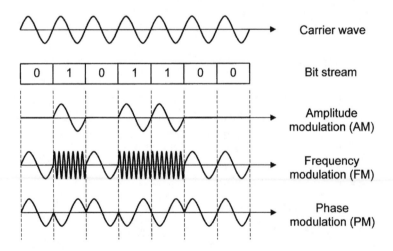

Figure 9.12: Three types of modulation, AM, FM, and PM, are used to put the digital bit stream onto the carrier wave.

contained in the digital bit stream onto a *carrier wave*, which can then be transmitted to the satellite. The carrier wave's job is simply to carry the information from ground to spacecraft. Initially, a carrier wave is an EM wave with a single frequency, as shown at the top of the figure. The next layer of the diagram shows the digital bit stream that needs to be carried by the wave, and there are three main ways of doing this. The first is *amplitude modulation* (AM), which is shown in the third layer in the figure. The amplitude of the wave is varied depending on the value of the bit; in this case, the amplitude is zero when a 0 bit is being carried, and nonzero when a 1 bit is transmitted. For *frequency modulation* (FM), as shown in the fourth layer of the diagram, the wave frequency is varied. When a 0 is carried, the frequency is low (long wavelength), and when a 1 is carried the frequency is high (short wavelength). Most conventional radios have AM and FM bands on their tuning dial. The third type, *phase modulation* (PM), is shown by the bottom line of the diagram. In this case, every time the value of a bit changes from a 0 to a 1, or vice versa, the phase of the carrier wave is changed by 180 degrees (see information on phase in Figure 9.10). In digital space communications, this is the most commonly used form of modulation.

When the carrier wave is received at the destination ground station, the process described above needs to be reversed so that the person at the other end of the phone can hear what the speaker said. The bit stream needs to be recovered by a process of *demodulation* of the carrier wave, and the analogue signal—the voice current—needs to be recovered by *decoding* the digital bit stream. It does seem to be a complex process, but it is all done routinely everyday, without anyone noticing!

It is also important to realize that there are thousands of these telephone conversations going on at the same moment, all of them sharing the same uplink to the spacecraft. To prevent one telephone conversation from bumping into another, each has a separate carrier wave at slightly different frequencies. So typically the ground station transmits a great wodge of carrier waves with frequencies ranging from, say, 6 to 6.5 GHz. On receipt of the uplink signal, the job of the satellite is to amplify the signal, change its frequency to, say, 4 to 4.5 GHz, and then retransmit it on the downlink to the destination ground station. The amplification is necessary as the signal strength after the uplink journey is small, about 10^{-8} W (or 0.000 000 01 W), a tiny fraction of the output of a domestic light bulb! The change in frequency is needed because the spacecraft usually uses the same antenna for both uplink and downlink, and the frequency shift prevents the information in the two links from becoming scrambled.

The Communications Subsystem

The focus of our discussion so far has been on telephone communications, but space communication is not just about this. For example, images taken by interplanetary spacecraft or Earth-observation satellites are commonly conveyed by space communications links, and this is again a digital process. The cameras on spacecraft use digital photography, so the pictures already come in a handy digital form, allowing the same type of method that we described above to transmit them from the spacecraft to the desktop computers on the ground.

The design of the spacecraft's communications subsystem is strongly influenced by the ground communications systems that it will talk to during the mission. Figure 9.13 shows typical spacecraft and ground communications antennas, used by GEO communication satellites. The main job of the communications subsystem engineer is to determine the size of the spacecraft antenna and the amount of power it needs to radiate to achieve a good-quality link. The physical characteristics of the overall (spacecraft and ground) system play a major part in this design process, and these include the ground-to-spacecraft range, the frequencies used, the size and power of ground antennas, and the losses caused by atmospheric absorption.

Figure 9.13: Spacecraft and ground communications antennas are shown, typical of those used for a GEO communication satellite mission. The ground station picture was taken at the Goonhilly Down ground station during a family holiday in Cornwall. (Spacecraft image courtesy of Lockheed Martin.)

The two most important attributes of the spacecraft communications subsystem are *radiated power* and *gain*. If you use a mobile phone, you have some idea of what the radiated power of a communications system is all about; they have a little scale indicator on the screen to tell the user if there is a signal. If there is one, then somewhere nearby there is a mobile phone mast that has been designed to have sufficient radiated power to cover a particular area so that calls can be made and received. Further away from the mast, the signal strength will drop, so another mast has to provide ground coverage area there so that the service can be maintained. In the same way, an important measure of the effectiveness of a spacecraft's communications subsystem is the amount of radiated power it generates. However, a number of watts of radiated power is not the end of the story; the effectiveness of the system can be further enhanced if the spacecraft's communications antenna (usually dish-shaped) has high gain. The simple rule is that a large dish has high gain, whereas a small dish has low gain. We can understand the idea of gain if we consider the small light bulb inside a flashlight. The amount of radiated power it generates, in terms of the amount of light it gives off, is small, and certainly much smaller than that of a standard domestic light bulb. If we take the small bulb out of the flashlight and connect it to a battery, it provides insufficient light in a darkened room. However, if we put the bulb back into the flashlight and turn it on, the flashlight's dish-shaped reflector focuses the light from the small bulb into a beam, which can be dazzling if pointed directly into your eyes. The flashlight's reflector has a measure of gain, effectively increasing the radiated power of the bulb along the axis of the flashlight where the beam is concentrated.

This is very much like a dish antenna on a spacecraft. The radiated microwave power generated by the spacecraft is focused into a beam by the dish, and this beam can then be pointed at a receiving dish on the ground. This has the effect of increasing the received power on the ground. To achieve the level of received power on the ground required to ensure a good quality of link, spacecraft designers have a choice: they can achieve the required level either by having a small amount of radiated power and a large gain (a big dish), or by having lots of radiated power and a small gain (a small dish). This *power-gain tradeoff* is one of the main design issues for the spacecraft communications subsystem engineer. This issue affects space-craft design, in particular that of interplanetary spacecraft that travel to distant parts of the solar system. For example, probes such as Cassini/ Huygens (see Fig. 7.6 in Chapter 7), and the New Horizons spacecraft (Fig. 9.14) recently launched to investigate Pluto, look almost like flying communications dishes. At great distances from the Sun, generating large amounts of electrical power onboard the spacecraft is difficult, so the

Figure 9.14: An artist's impression of the New Horizons spacecraft at Pluto. (Image courtesy of NASA.)

communications subsystem is likely to have low radiated power. To achieve the link quality required to return science data across such great distances, these spacecraft require large, high gain antennas.

Let's move on now and briefly discuss the subsystem that is tasked with the job of moving all the binary digits around the spacecraft.

The On-Board Data Handling (OBDH) Subsystem

As we recall from Chapter 7, the main functions of the OBDH subsystem are to provide storage and processing of payload and other data, and to allow the exchange of data between subsystem elements. As this brief description suggests, the subsystem is made up mostly of computers, their peripherals, and software. The OBDH subsystem is distributed throughout the vehicle, to make possible the data links between all the spacecraft subsystems needed for successful operation, such as the ACS control loop we discussed in Chapter 8 (see Figure 8.1), and the transfer of payload data from the payload equipment to the communications subsystem ready for downlink to the ground. Although the OBDH subsystem is a virtual one, in the sense that it consists mostly of computer processors and programs, nevertheless it is perhaps the most real part of the spacecraft for the operators as it is the part that they talk to. This interaction with the ground takes place in two directions—through the uplink, which is mostly *command*, and through the downlink, which is mostly *payload data* and *telemetry*.

The Command Function

It is through the command uplink that the ground operators talk to the spacecraft to get it to do something. Commands are received, interpreted, and distributed by the OBDH subsystem. The commanded tasks may be as simple as, say, switching a battery heater on, or may involve a more complicated job, such as repointing a space observatory to a new place in the sky, or commanding an Earth-observation satellite to image a particular ground target of interest. This command function is important, as it allows the ground team to control all aspects of the spacecraft's operation. As such, it must be done in a reliable way; for example, the uplinked commands must be verified, to ensure that they have been received correctly. The OBDH subsystem must then provide confirmation that the uplinked command instructions have been carried out correctly, which is usually done through the use of telemetry (see below). All this sounds rather tedious, but some very expensive spacecraft missions have been lost simply due to incorrect or unverified commands being executed on board the spacecraft.

The Payload Data and Telemetry Functions

Perhaps the most important role of the OBDH is to ensure that data generated by the payload are transferred to the communications subsystem ready for downlinking to the ground. In some cases, these payload data need to be processed by the OBDH subsystem computer(s), and this processing involves things like *storage, error detection and correction* and *compression* (see below). For some spacecraft, such as communication satellites or those with imaging payloads, the volume of data produced by the payload can be large. For LEO spacecraft, sometimes a ground station is not in view as the payload data are being generated, and so the data must be stored on board while waiting for an opportunity to downlink to a ground station. These storage devices are part of the OBDH subsystem, and in the past tape recorders were commonly used. More recently, *solid-state memories*, like those in a desktop computer, have been used instead. They provide huge amounts of data storage, up to hundreds of Gbits (where the "G" stands for "Giga," meaning 1000 million), but they are prone to radiation-induced errors, as mentioned in Chapter 6. To minimize the effects of such data errors, the memory is continuously scanned by error detection and correction computer programs, as part of the OBDH subsystem function. In some cases, the data volume generated by the payload is too large for the downlink communications to handle, and so the data need to be compressed using OBDH software. The raw payload data can be compressed (or reduced in volume) in such a way as not to compromise the value of its content too much. This is done by eliminating redundant or duplicated content, removing unwanted information, or reducing the resolution of digital imagery.

The other main job of the OBDH subsystem is to generate telemetry, which we discussed briefly in Chapter 7. Various sensors are placed around the spacecraft, to monitor the health and operating status of the onboard equipment. They check the temperatures of electronic equipment, the pressures in fuel tanks, voltages and currents in power supplies, and so on. The operational status of equipment is also monitored in terms of whether items are switched on or off. All these accumulated data are converted into a digital bit stream, which is then downlinked with a data rate of a few kbps (kilobits per second) to the ground for display on computer monitors in the operations room. In this way, any problems that occur on the spacecraft can be quickly spotted and corrected by the operations team.

The Thermal Control Subsystem

For the majority of spacecraft in orbit, the thermal control subsystem represents only a small percentage of the spacecraft mass, yet remarkably it generally dominates the overall appearance of a spacecraft. It is what you see! To illustrate the point, Figure 9.15 shows Earth-observation spacecraft SPOT 5 prior to launch (see Table 7.4 in Chapter 4). It looks a bit like a chocolate box wrapped with gold and silver foil, and this "foil" is the *multilayered insulation blanket* that I have mentioned briefly in previous chapters. There are also other features visible, such as mirror-like surfaces and white-painted areas and antennas, and these are all characteristic of a typical thermal control subsystem design. The same features can be seen in many of the pictures of spacecraft in earlier chapters, with the thermal design dominating their appearance. In this section, we have a look at why this is.

As we recall from Table 7.1, the main job of the thermal control subsystem is to provide an appropriate thermal environment onboard, to ensure reliable operation of payload and subsystem elements.

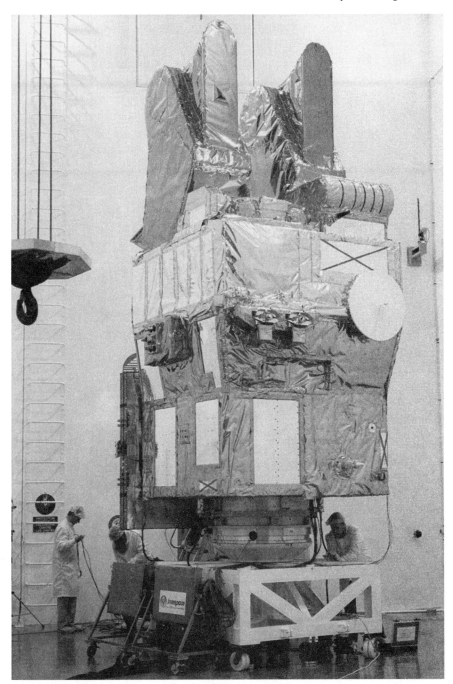

Figure 9.15: The SPOT 5 Earth-observation satellite. The thermal control subsystem design governs the way the spacecraft looks. (Image copyright © CNES/Patrick Dumas.)

Table 9.3: Approximate temperature range for reliable operation of equipment

Component	Approximate temperature range (°C)
Batteries	0 to 25
Propellant (e.g., Hydrazine)	10 to 50
Electronic equipment (e.g., computer processors)	−5 to 40
Mechanical bearings (e.g., reaction wheels)	0 to 45

Equipment Reliability

The average spacecraft is crammed with electronic and mechanical equipment so that it can achieve its mission. Most of this equipment, especially the electronics, has been developed from similar domestic and industrial items that were designed to operate down here on the ground. In other words, they have a *design heritage* such that they work best when they operate at room temperature. One main reason for requiring thermal control onboard a spacecraft is to produce a room temperature environment for the various equipments so that they operate reliably for the lifetime of the spacecraft mission. It is the same with home electronics. For example, if we put our TV set in the freezer or in the oven, its useful lifetime will be considerably reduced. Similarly, reliable, long-term performance of most spacecraft components requires them to operate within thermal tolerances. Table 9.3 shows the approximate temperature range for the reliable operation of some items of spacecraft equipment, which the thermal control engineer has to consider when designing the thermal control subsystem. (The definition of room temperature is quite broad.)

Payload Requirements

As we have already mentioned in Chapter 7, certain payloads may have to be maintained within a strict temperature range to work properly, which will affect how the thermal control subsystem is designed (see Figure 7.1). To illustrate, we can use the example of a spacecraft that has a large imaging payload on board, such as an Earth-observation satellite or a space observatory. In both types of spacecraft the imaging equipment is made up of a variety of mirrors and lenses, and their job is to bring the light entering the payload to a focus in the right place so that the image can be recorded in a way similar to that done by a digital camera. To get the image in focus, the lenses, mirrors, and detectors all need to be kept at the right distance from one another. To do this, the optical components are all firmly mounted on a rigid framework, usually referred to as an *optical bench*, within the spacecraft. This framework needs to be a robust structure so that the imager still works after the rough ride to orbit on the launch vehicle (see Chapter 5). But another feature of importance is its thermal design. When in orbit, the spacecraft is exposed to extremes of temperature, and if unprotected from these, the optical bench framework will expand and contract in size in response to changes in temperature. From the point of view of the image quality, this small, relative movement among the mirrors, lenses, and detectors is clearly undesirable. Consequently, it is the job of the thermal control engineer to design the spacecraft so that the payload is isolated from these extremes of temperature, in order to ensure that it works. In large observatories, such as the Hubble Space Telescope (see Table 7.5 for details), this job can be challenging.

The In-Orbit Thermal Environment

To keep the temperature of the spacecraft and its components within the required range, the thermal control engineer has to consider the factors that heat the spacecraft and those that cool it. The engineer strives to achieve a balance, so that the spacecraft does not get too hot or too cold. If we focus on a spacecraft in Earth orbit, we can summarize its thermal environment in terms of heat inputs (factors that heat it) and heat outputs (factors that cool it).

Heat Inputs

The mechanisms that heat the spacecraft are shown in Figure 9.16. As we mentioned previously in Chapter 6, the major input is that due to direct solar radiation, that is, the direct electromagnetic radiation from the Sun. This amounts to about 1.4 kW of thermal power for every square meter of spacecraft area that is presented to the Sun. In addition, the spacecraft receives solar radiation that is reflected from Earth's surface. About one third of all the sunlight that falls on Earth is reflected back out into space, mostly from cloud and ocean surfaces. This is referred to as *Earth albedo radiation*, and this too heats the spacecraft. Another source of heat, referred to as *Earth heat radiation*, is the infrared radiation (see Figure 6.2) produced by Earth simply because it is a warm body. Sometimes we are very much aware of heat (infrared) radiation, such as when we sit in front of a glowing open fire and feel the heat on our faces. However, such heat radiation is produced by all objects to a greater or lesser extent depending on their temperature. This is true as long as the object's temperature is above *absolute zero*. On the Celsius temperature scale, absolute zero occurs at –273° and is so called because it is

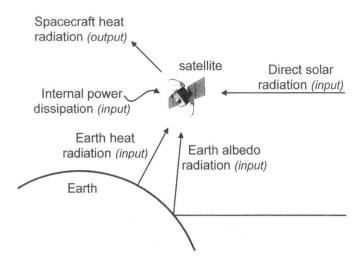

Figure 9.16: A summary of the spacecraft's thermal environment, in terms of things heating it (inputs) and things cooling it (outputs).

the lowest temperature that is physically possible. At $-273°$C all physical processes stop.

Objects such as Earth and people give off heat radiation simply because their temperature is above absolute zero. Since Earth is an object with a temperature of around 20°C (on average), it produces a heat radiation field that gently warms the spacecraft. Both of these Earth-generated effects, Earth albedo and Earth heat radiation, provide a heat input to the spacecraft that decreases with the altitude of the spacecraft's orbit in accord with an inverse square law (see Chapter 1). The final mechanism that heats the spacecraft is not an external one but rather an internal effect referred to as *internal power dissipation*. Typically a spacecraft is packed with electrical and electronic equipment, and most of it is not very efficient in that a significant percentage (between 10% and 50%) of the electrical power that is fed into the equipment to sustain its operation is wasted in producing heat. This is not just a feature of spacecraft components; the same thing happens in domestic electrical equipment. For example, if we leave the television set on for a couple of hours, and then just place our hand (carefully) by the air vents at the back of the set, we can feel that not all the electricity has been used to produce picture and sound. Some of it is being wastefully dissipated as heat. Internally dissipated power is another significant mechanism that acts to heat up the spacecraft.

Heat Outputs

Unless there is some way of getting rid of the heat, the spacecraft is going to become too hot, and the acceptable temperature ranges shown in Table 9.3 will be exceeded. However, as a warm body, the spacecraft itself will give off its own heat radiation, and the intensity of this radiation will increase as the spacecraft's temperature rises. This spacecraft heat radiation (see Figure 9.16) is the only effective thermal output acting to cool the spacecraft down.

Thermal Equilibrium

From the above discussion about the thermal environment, we can see that at a particular temperature, the heat outputs will match the heat inputs, and a kind of thermal equilibrium will be reached where the temperature of the spacecraft remains more or less constant. This temperature is referred to as the *equilibrium temperature*. It is the job of the thermal control engineer to design the thermal control subsystem to ensure that when thermal equilibrium occurs, the equilibrium temperature is around room temperature. If this can be achieved, then there is a good chance that all of the on-board components will be able to operate reliably for the lifetime of the mission.

Thermal Control Design

How does the thermal control engineer achieve this objective? In the introduction to this section we said that the thermal control subsystem is what you see, and basically this is the clue to how it is done. First we need to discuss the thermal properties of materials, that is, why some surfaces get hot in the Sun, while others stay cool. If we walk barefoot on a beach on a hot summer's day, some surfaces, such as the sand or black tarmac on the seafront promenade, get hot and scold the soles of your feet, whereas others, such as the wooden steps down to the beach, feel quite comfortable.

Along the same theme, I have a friend who used to work as a spacecraft thermal control subsystem engineer, and he had a neighbor who owned a large motor boat. But there was a problem with the boat: parts of the decking were made of stainless steel, and when exposed to the Sun on a hot weekend the temperature of the decking would rise to a level that was hazardous. My friend did a few simple calculations and estimated that the temperature of the stainless steel decking could reach temperatures in excess of 100°C! The solution was an easy one; my friend recommended that the offending parts of the deck be painted white. As we will see below, a white-painted surface is poor at absorbing the Sun's heat but good at giving off heat radiation. As a consequence the temperature of the deck went down to a comfortable room temperature, and the boatman was happy. Obviously, boat builders can learn something from spacecraft engineers!

Some of the spacecraft's surfaces are good at absorbing the Sun's heat, but not good at giving off heat (infrared) radiation when they get hot. An example of this kind of surface is aluminium; it is quite good at absorbing the Sun's heat, but not good at radiating it. An aluminium surface exposed to the Sun in Earth orbit can reach temperatures of around 300° to 400°C. On the other hand, some surfaces are poor at absorbing the Sun's heat, but are good at giving off heat radiation, such as a white-painted surface, which stays cool even when exposed to direct sunlight. If we apply a coat of white paint to our orbiting aluminium surface, its temperature will drop to around, say, 20°C. Obviously, the precise numbers here depend on the orbit, which determines how long the spacecraft is in direct Sun, and how long it is in Earth's shadow. These so-called thermal properties of the surface materials of the spacecraft are used to good effect by the thermal control subsystem engineer to manage the balance between heat inputs and outputs to ensure acceptable local temperatures and an appropriate overall equilibrium temperature—around room temperature—for the spacecraft. This is why spacecraft look the way they do, with various types of surface material to manage this thermal balance.

In Figure 9.15 we can see a white-painted communications dish to ensure it stays cool in direct sun light. However, the main feature of the thermal design in this case is the extensive use of thermal blanket to insulate the spacecraft from the direct heating effect of solar radiation. This often gives the spacecraft its characteristic appearance of being wrapped in gold or silver foil. The blanket, sometimes referred to as *multilayered insulation* (MLI), is composed of multiple layers of a thin plastic film with a metallic coating of aluminium, silver, or gold. Each individual layer is similar to the survival blankets handed out at the end of marathons to keep the runners warm. Figure 9.17 shows a small section of MLI, which is made up of around 25 such layers. Each layer is perforated with tiny holes, to ensure that the air trapped in the blanket on the ground can escape easily when the spacecraft reaches the vacuum of orbit. Each layer of the MLI is separated from the next by a thin sheet of nylon bridal veil. In orbit, with a vacuum between each separate layer, the effectiveness of the insulation provided by the blanket is maximized.

However, if we were to wrap the spacecraft completely in multilayered insulation, the heat dissipated by the electrical equipment on board would not be able to escape easily, and the interior of the spacecraft would become hot. So sections of MLI in Figure 9.15 are cut away, and instead radiator surfaces are installed, which are usually mirror-like surfaces that are designed to be poor absorbers of the Sun's heat but good at giving off heat (infrared) radiation. As a result they remain cool, even in direct Sun, and encourage the escape of internally dissipated heat. Usually, the electrical

Figure 9.17: A section of multilayered insulation, with one layer folded back to reveal the nylon bridal veil spacer. Each layer is perforated with small holes, spaced at approximately 1-cm intervals, to allow trapped air to escape from the layers.

Figure 9.18: A lesson from nature in thermal control: polar bears displaying their insulation blankets and thermal radiator surfaces.

devices that produce the most heat are mounted on the interior surfaces of these radiators to keep them cool.

Like many aspects of engineering, this arrangement is a lesson from nature, and is similar to the way a polar bear, for example, manages its thermal control! A polar bear (Fig. 9.18), like a spacecraft, has to survive in a hostile thermal environment. To keep warm in the cold polar regions, it needs good insulation in the form of its fur coat. On the other hand, it does need some effective radiator surfaces, so that it does not overheat when it exerts itself physically or when summer comes. So it has radiators too, such as the pads on its feet, its shiny black nose, and its tongue, to allow its internal heat to be radiated away. Like the spacecraft, it must have enough insulation to keep warm when its environment gets cold, but enough radiator area to cool down when the environment gets too hot.

The Structure Subsystem

The function of the spacecraft structure is to provide a rigid framework to carry all the various payload and subsystem equipment. As we recall from Table 7.1, its function is to provide structural support for all payload and subsystem hardware in all predicted environments (especially the harsh launch vehicle environment). The important phrase here is "in all predicted environments." When structure subsystem engineers begins the process of designing a new spacecraft, the first thing they look at is the most severe of these environments—that of the launch vehicle. In Chapter 5, we saw how harsh this is, with high levels of acceleration, vibration, shock, and noise. The launch vehicle agencies provide detailed information on all these aspects of the ride to orbit, and the main job of the structure subsystem engineer is to produce a structural design that will survive the journey from launch pad to orbit.

Design Requirements
The following is a list of the important aspects that the structure subsystem engineer has to consider, many of which are related to the fundamental requirement to survive launch:

- Low mass. Although there is a need to build a robust structure, nevertheless the structure subsystem engineer must also make every effort to minimize mass. As we said in Chapter 5, the cost of the launch rises steeply as the mass of the spacecraft increases, so this requirement becomes a critical issue about limiting the overall cost of the spacecraft project, as launch is usually a large percentage of this cost.

- Strength and stiffness. The structure must be strong and stiff enough to withstand launch and on-orbit loads without distortion. Unacceptable levels of distortion would compromise the pointing of payload instruments, such as imaging cameras or telescopes, and the pointing of subsystem elements like communications dishes and attitude sensors.
- Environmental protection. The structure must also provide an appropriate level of protection against environmental aspects (see Chapter 6), such as radiation, and shielding against impacts of orbital debris and micrometeors.
- Launch vehicle interface. The manner in which the spacecraft is attached to the launch vehicle also affects the spacecraft's overall design. The attachment must be done in such a way as to ensure a robust connection with the launcher during the ascent to orbit, but it must also be able to reliably release the spacecraft on command once in orbit. The position of this interface within the spacecraft also governs how the launch loads are distributed throughout the spacecraft structure.

Materials

So what kinds of material are used to satisfy the demanding requirements related to having a robust structure, but nevertheless one that is of low mass? A material that is commonly used in spacecraft construction currently is aluminium honeycomb panel. It is composed of a sheet of aluminium honeycomb, bonded between two thin sheets (skins) of aluminium, as shown in Figure 9.19. The honeycomb is basically identical to bees'

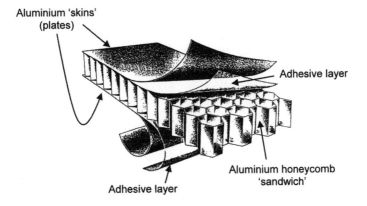

Figure 9.19: Aluminium honeycomb panel is a commonly used material in spacecraft manufacture. Thin aluminium skins are bonded onto a sandwich filling of aluminium honeycomb, making a very light, stiff panel.

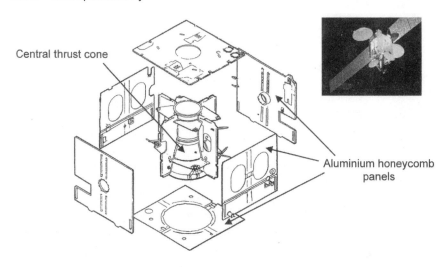

Central thrust cone

Aluminium honeycomb
panels

Figure 9.20: An example of a spacecraft structure, based on the Eurostar®
configuration. The figure shows the central box-like structure, made up of a number
of honeycomb panels. The thrust cone is the housing for the spacecraft's primary
propulsion, and is also the position of the interface with the launch vehicle. (Images
courtesy of EADS Astrium.)

honeycomb, but made of aluminium. A sheet of this seems fragile and floppy
when handled, and can be easily crushed with a pair of pliers. However,
when the aluminium skins are glued either side, making a honeycomb
sandwich, the resulting panel material is very light and stiff, providing an
ideal material for spacecraft manufacture.

An example of the use of this material is given in Figure 9.20, which shows
the basic structure of the central box-like body of a communications
satellite. The central thrust cone is the housing for the spacecraft's primary
propulsion system, and also the location of the interface with the launcher.
As such, this element takes most of the loads during the launch and when the
main rocket engine is fired on orbit.

Summary

Figure 9.21 is an exploded view of a spacecraft that displays many of the
subsystem elements that were discussed here, thus providing a brief
summary of the contents of the last three chapters.

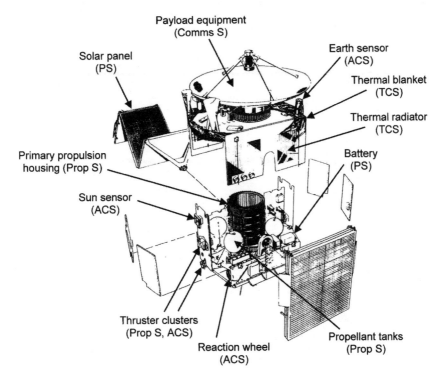

Payload equipment
(Comms S)

Earth sensor
(ACS)

Solar panel
(PS)

Thermal blanket
(TCS)

Thermal radiator
(TCS)

Primary propulsion
housing (Prop S)

Battery
(PS)

Sun sensor
(ACS)

Thruster clusters
(Prop S, ACS)

Reaction wheel
(ACS)

Propellant tanks
(Prop S)

Figure 9.21: An exploded view of a spacecraft showing the various elements of the subsystem design. The example shown is that of a communications satellite, so in this case the payload is identified with the communications subsystem.

Key: ACS, attitude control subsystem; Comms S, communications subsystem; PS, power subsystem; Prop S, propulsion subsystem; TCS, thermal control subsystem. (Backdrop image courtesy of ESA.)

Space in the 21st Century

SOMETIMES as a child, I engaged in the rather pointless activity of wishing I had been born later! This was in the 1950s, when the exploration of the moon and the solar system was yet to begin, and I was impatient to know whether there were little green men on Mars, and what the ringed majesty of Saturn would look like above the horizon of its moon Titan. In this, my imagination was nourished by many evocative pictures by space artists, such as the iconic image in Figure 10.1 by Chesley Bonestell. We now know that Titan has a thick, murky atmosphere, so that such a vista is sadly unlikely to be echoed in reality. I have to admit that there is still something of this childish outlook in me now, and I am still impatient to know how the fundamentals of the universe work, and what it would be like to be able to actually see the sights of the solar system for myself, such as the rings of Saturn or the volcanoes of Jupiter's moon Io.

No one knows what the future holds, but it is possible to imagine a time when physicists will have come up with a *theory of everything* to explain how the universe works, and when the engineers will have solved the problem of interplanetary, interstellar, and maybe even intergalactic travel. My feeling is that there really is no reason to believe we cannot go faster than the speed of light—a speed limit imposed upon us by Einstein's physics. Why should Einstein's theories be the last word in physics, in the same way that Newton's theories seemed unchallengeable at the turn of the 20th century?

Although I can get rather excited about these future prospects, at the end of the day I just have to calm down and accept my allotted position in space and time. From where I am at the moment, it seems unlikely that I will see a fundamental breakthrough in physics that will allow us to understand everything about the universe we inhabit, and how it all works. Coming a little closer to home, if I am lucky I might get to see the first human set foot on the planet Mars. It is interesting to think that, whoever they are, they are almost certainly alive today as I write. And coming a little closer still, I am fairly optimistic that I will see people on the Moon again within the next

G. Swinerd, *How Spacecraft Fly: Spaceflight Without Formulae*,
DOI: 10.1007/978-0-387-76572-3_10, © Praxis Publishing, Ltd. 2008

Figure 10.1: Saturn as seen from Titan, depicted by Chesley Bonestell. (Image courtesy of Bonestell Space Art.)

decade or two. I suppose my impatience stems from the fact that progress does seem so painfully slow; after all, in cosmic terms the Moon is just on our doorstep!

This is not, however, to undermine the achievement of the Apollo program. Some people would say that the late 1960s and early 1970s, when the Apollo astronauts stood on the Moon, was the golden age of astronautics. A team of bold young men—and maybe not so young as far as the astronauts themselves were concerned—made John Kennedy's vision of 1961 happen. Figure 10.2 shows the lunar lander of Apollo 12 on the plain at Oceanus Procellarum. At the time of the Apollo program, I was a teenager and completely enthralled by the whole business! In fact, Apollo is probably one of the reasons why I have spent my career working, teaching, and researching in the area of space. Apollo was inspirational! At the time, with my young idealistic view of the world, I believed that the Americans went to the Moon with the best of intentions—to further our knowledge of this little corner of the universe. But with the benefit of hindsight, and imbued perhaps with a dollop of cynicism that comes with age, I now can appreciate that the main reason for doing it was for capitalism to demonstrate its superiority over Communism. Twelve years before the first moon landing, the launch of the first satellite by the Soviet Union, Sputnik 1, had severely dented American pride, and Apollo was a way of redressing the balance. Despite this, however, Apollo was an outstanding achievement from a space engineering point of view—to say nothing of the courage of the men who actually stood on the surface of the moon. And nothing like it has been seen since. If you had told me in 1972, when Eugene Cernan and Harrison Schmitt took off from the Moon as the final act of the Apollo program, that no one would have returned in 35 years, I would not have believed you! This really does emphasize the one-shot nature of the way it was done as primarily a political act.

If Apollo was the golden age, then unfortunately most of the young people today have missed it. Although many good things have happened in spaceflight since, as we have discussed in this book, nevertheless you could argue that there is currently nothing to inspire young people to get involved in space engineering. Clearly the emphasis in the interim has shifted from space exploration to space applications. This move to use space for communications, navigation, and Earth observation has revolutionized the business and leisure worlds, but it is perhaps not quite so inspirational as, say, opening the solar system to manned space exploration. There is a great need to do something that will inspire young people to get involved. It really is only the young that can dream the big dreams, and make them happen!

In this chapter (which is again longer than average) and the next, I will attempt to discuss some of the developments we may see in the future.

Figure 10.2: The lunar module of Apollo 12 landed in Oceanus Procellarum (the Ocean of Storms) in November 1969. This was the second moon landing of six. When will we see the like of this again? (Image courtesy of the National Aeronautics and Space Administration [NASA].)

Manned Spacecraft

Before we begin this rather speculative journey, we need to briefly discuss manned spacecraft. At this point, I should perhaps repeat the caveat from Chapter 2 about the use of the phrase *manned spaceflight* to mean flights involving both men and women. I know that the phrase may not be quite politically correct, but I dislike the other possibilities, such as "crewed" missions or "peopled" missions.

In earlier chapters we discussed how unmanned satellites are designed, and now we discuss the additional design requirements for manned spacecraft. Space is a hostile environment (Chapter 6), and people are fragile organisms, requiring air to breathe at the right pressure, food to eat and water to drink, as well as an acceptable ambient temperature and a way to get rid of personal waste. These rather obvious requirements translate into the need for a significant mass of hardware and provisions onboard manned vehicles. This trend is readily apparent when we consider current examples of manned space vehicles such as the International Space Station (ISS) (Fig. 10.3), which is

Figure 10.3: The International Space Station (ISS) as it is projected to look when completed in 2010. (Image courtesy of NASA.)

projected to have a mass on the order of 450 metric tonnes when its construction is complete around 2010. Another feature that tends to further increase the mass of manned spacecraft, which is perhaps less obvious, is that of *redundancy*, which entails having backup systems onboard to make the vehicle safe from failures that would threaten the lives of the occupants. Safety-critical items such as elements of the life support system are doubled-up so that if the primary element fails, the backup system can be brought online to ensure the well-being of the crew. This is a major issue in manned vehicle design; a line has to be drawn by the design engineers to ensure a balance between having an excessively massive and extremely safe spacecraft on the one hand, and having a less massive but potentially unsafe vehicle on the other. From the point of view of launch costs, the less massive option is favored. But this is certainly not the case from the point of view of crew safety!

So manned spacecraft are expensive. Not only is the spacecraft hardware to be lifted to orbit more massive, due to the need for life support systems, but the launcher has to be *man-rated*. As we saw in Chapter 5, this means that the launch failure rate has to be reduced from the typical 10% for unmanned launchers to something significantly less than 1% for the man-rated launch vehicle. This is done by increasing the amount of redundancy in the launcher itself, which translates into more mass. Good examples of man-rated launchers, which carry manned hardware to orbit, are the Saturn 5, which sent the Apollo hardware and astronauts on their way to the Moon, and the Space Shuttle. As a consequence, these launch vehicles are some of the most complex, massive, and expensive launch systems yet to be employed. As we look to the future, and the projected expansion of manned space exploration and possibly space tourism, this issue of the cost of access to orbit is one of the main stumbling blocks that will ultimately need to be removed. Some insights into the challenges this poses for rocket scientists were discussed in Chapter 5.

The impact of including people in the design of spacecraft entails not only emulating a suitable terrestrial environment onboard to support life, but also dealing with the aspects of the space environment that are hostile to human life, such as radiation, space debris, and microgravity. The following factors influence how manned spacecraft are designed and operated (also see Chapter 6):

Radiation

In terms of radiation, the main threat to the health of people in space comes from particle radiation (as opposed to electromagnetic radiation). As we saw in Chapter 6, this is essentially made up of energetic (rapidly moving) subatomic particles, such as electrons, protons, and sometimes ions (the nuclei of atoms stripped of their attendant electrons). This type of radiation comes from a number of sources. For people in Earth-orbiting spacecraft, the main offending source is the Van Allen radiation belt, which contains high-energy electrons and protons that have been captured from the Sun and trapped by Earth's magnetic field. However, in low Earth orbits of altitudes less than about 1000 km (620 miles), spacecraft are well below the most intense parts of the Van Allen belt, and are also protected from the majority of direct solar particle radiation by the Earth's magnetic field. As a consequence, astronauts living long-term in the ISS, at an altitude of around 350 km (220 miles), suffer a relatively low level of potentially damaging radiation. However, the use of higher orbits for long-term habitation, for example in the most intense part of the proton radiation belts, which is at a height of about 4500 km (2800 miles), would result in a fatal radiation dose for the crew.

Once the spacecraft leaves Earth orbit and the shelter of Earth's magnetic field, we have another situation entirely. Future trips to other planets will involve astronauts traversing large distances, taking hundreds of days to reach their destination. In these cases, the crew is at the mercy of direct particle radiation from the Sun. At times of solar maximum, solar storms can occur that fling huge quantities of particle radiation across the solar system. If the spacecraft happens to be in the path of one of these outbursts, the level of radiation can be potentially lethal for an unprotected crew. Thus the problem of providing adequate radiation protection would appear to be a potential roadblock for future manned flights to the planets. However, there is a cost-effective two-part solution. First, there must be an effective early warning system that monitors the Sun's output to detect the solar storms, probably using a system of spacecraft sensors in orbit around the Sun. A storm warning can then be communicated to the distant manned spacecraft. Second, there must be a *storm shelter,* which is a small pressurized compartment onboard the manned vehicle where the crew can stay during the solar storm. To protect the crew, you might expect that this shelter needs to be lined with a considerable thickness of lead. However, a better solution, in terms of reducing mass, is to reorient the spacecraft to put a significant mass of existing spacecraft hardware or propellant between the shelter and the Sun. For example, it is estimated that about a half-meter thickness of liquid hydrogen propellant would provide adequate radiation protection for the crew.

Space Debris

This is a risk mainly for manned spacecraft in low Earth orbit (LEO), where the amount of space junk is sufficient to cause concern that a damaging collision may occur. Manned spacecraft that operate long-term in the LEO environment, such as space stations, can be equipped with debris bumper shields (see discussion of Whipple shields in Chapter 6) to protect them from debris impact. For example, a considerable amount of design effort and mass has been invested in the ISS to provide debris shielding. Most of this shielding is deployed on the forward-facing surfaces of the station, as most damaging impacts are likely to be caused by debris coming from the forward flight direction. On the other hand, the Space Shuttle is an example of a manned vehicle that cannot be equipped with such shielding, due to the fact that it has to fly both in the environment of space and in the Earth's atmosphere. The use of Whipple-type shielding would compromise its ability to fly through atmospheric reentry and landing. Over recent years, there has been a growing appreciation of the threat from debris impact to the Shuttle orbiter and its crew. In an attempt to at least minimize the threat to the crew, the vehicle adopts a particular attitude in orbit—upside-down and with the main engines facing forward. In this way the crew is protected from impact with the potentially most damaging debris coming from the forward direction.

Microgravity

Microgravity, or weightlessness, in orbit has been an issue not for the design of manned spacecraft but rather for the effect it has on the physiology of the human occupants. Trying to determine and understand the effects of long-term weightlessness on people has been a major preoccupation of manned space programs since the first orbital flight of Yuri Gagarin in 1961. Considerable medical data have been gathered over the years, with astronauts aboard space stations such as Skylab, Salyut, Mir, and now the ISS staying in orbit for hundreds of days and acting as willing guinea pigs. This means that the effects of microgravity over a period of time typical of, say, a manned flight to Mars can be studied and evaluated. The main physiological effects of weightlessness on people can be summarized as follows:

- Motion sickness: The balance sensors we have in the inner ear rely on the movement of fluid in a normal $1g$ environment to give us information about how we are oriented (lying down or upside-down, for example) and how we are moving around. With the fluid in a weightless condition, the brain has difficulty interpreting what the balance sensors are saying, and there is also a conflict between this information and what the eyes see, which can result in nausea and illness, with some astronauts being affected more than others. The brain usually takes about 2 or 3 days to sort out the new sensory inputs and to adapt to the new environment.

- Redistribution of bodily fluids: On Earth the blood pressure of a standing person decreases with the height above the feet; the pressure in the brain is about one third of that in the feet. Exposure to weightlessness causes a major redistribution of blood, resulting in a bloated face and thin legs—the classic so-called "puffy face and chicken legs" syndrome! The astronaut soon adapts to this situation in orbit, and the body appears to recover its normal function after a short period of time once back in a 1g environment on the ground.
- Muscle atrophy: Long periods of weightlessness usually mean physical inactivity, which causes muscles to waste away. This includes the heart muscle, which generally loses mass, with an accompanying reduction in heart rate. To combat this worrying trend, astronauts must spend a significant amount of time in necessary physical exercise, using complex exercise equipment designed for the microgravity environment.
- Bone decalcification: Another significant effect of weightlessness is the cumulative loss of calcium from bones, resulting in bone fragility in the long-term. This trend appears to be reversible once the astronaut is back in a 1g environment.

Although people seem to be able to recover from most of the effects of microgravity with time, the issues of loss of bone and muscle mass are a concern in future manned exploration of the solar system. Long-duration spaceflight in weightless conditions means the astronauts are clearly not in the best of physical condition when they arrive at their destination. Intensive physical exercise for the astronauts, and possibly the use of artificial gravity (see below) during the flight may be partial solutions to this problem.

Manned Space Exploration: The Immediate Future

This section discusses manned missions that are (almost) certain to happen in the next 30 years or so. The main elements of this vision are the following:

- The completion and operation of the ISS
- The resumption of manned exploration of the moon
- A manned mission to Mars

Despite a measure of age-induced cynicism regarding these objectives, I nevertheless have a reasonable level of confidence that they will happen. There does seem to be a degree of momentum behind these objectives, particularly with the January 2004 declaration by the incumbent U.S.

president of a new "Vision for Space Exploration". There are other signs as well, such as the retirement of the Space Shuttle fleet in around 2010. The Space Shuttle has been the workhorse of the U.S. space program since its first flight in 1981, and its retirement may at first seem to be a negative development. However, this will actually force changes in the U.S. space program, and encourage a wide-ranging rethink of future objectives and how they can be achieved. This new vision appears to be having a reinvigorating effect on the space program as a whole.

The other main issue is the huge cost of the projects listed above. As such, we can reasonably question the viability of pursuing these goals. In the days of the Apollo moon landings, the huge financial burden was justified by a political motive. But today, in the absence of the Cold War and the political competition that it created, it is reasonable to ask—What can justify and motivate nations and taxpayers to spend huge amounts of money returning to the Moon or going to Mars? It is easy to get immersed in the exciting technical aspects of this vision for the future, but there is no easy answer to this question, to which we will return later.

The International Space Station

At the time of this writing, manned space activity is almost entirely focused on the construction of the ISS in low Earth orbit (LEO). As the name implies, this is an international project involving the space agencies of the United States (NASA), Europe (ESA), Japan (JAXA), Canada (CSA), and Russia (RKA). Its orbit is a near-circular LEO with an inclination of 52 degrees, and a height of about 350 km (220 miles), although the altitude varies a little due to the effects of air drag. When completed in 2010, the mass of the station will be an impressive 450 metric tonnes, and Figure 10.3 shows its final configuration.

The construction began in 1998, and more than 40 assembly flights will have been required to complete the station's construction, the majority of these being Space Shuttle flights. Once completed, the largest overall dimension will be about 110 m (360 feet), the total available electrical power will be on the order of 100 kW, and the pressurized volume accessible to the six crew members will be around 1000 cubic meters (35,000 cubic feet). Beyond completion, the projected lifetime of the station is 6 years, so that it will be scheduled for a controlled de-orbit and atmospheric reentry around the year 2016.

Another impressive statistic is the anticipated cost of the project, about $130 billion, and it is this statistic that has drawn the most attention from the station's critics. The substance of most of this criticism is founded on the belief that the huge budget for the ISS could be better spent on unmanned spacecraft, such as observatories and interplanetary probes. The argument goes that the return in terms of science of unmanned exploration would be much greater, and it is easy to have sympathy for this point of view. The main science research goals of the ISS include astronomy and Earth observation, but we can see (from Chapter 2) that the ISS orbit is not ideal for either of these activities. Other research areas focus on experiments that require one or more of the unusual conditions, mostly related to microgravity, present on the station. These include the continued study of the effects of weightlessness on people, and studies in physics and chemistry, such as materials science. The argument between the two camps has been quite acrimonious at times. One outspoken opponent of manned space exploration is Bob Park, a professor of physics and formally the chair of the Department of Physics at the University of Maryland. His view is rather extreme, but nevertheless sums up the strength of feeling in opposition to the ISS among some American scientists. Park states, "NASA must complete the ISS so it can be dropped into the ocean on schedule in finished form"! (Park points out on his personal Web site that the views he expresses are his own and not those of the University of Maryland.)

My own feelings about the ISS are mixed. I have to admit that I have always thought it to be a very expensive project that is looking for a purpose to justify the cost. It can be argued that nations involved in programs like the ISS gain economic benefit through the development of a high-tech industrial sector, with the associated highly skilled work force. It is certainly the case that there are spin-offs from space technology development, and I am referring to more than just the much-quoted old chestnut—the Teflon frying pan! There is certainly economic benefit to be gained from space industry research and development spinning-off into commercial industry. But generally I would guess that the level of benefit in economic terms is very likely to be less than the investment. So I have to take an alternative tack in justifying expensive manned space programs like the ISS. The fact that the mission ends around 2016 also implies that the ISS will not be a part of the orbital infrastructure in aid of a return-to-the-Moon program or a manned mission to Mars. On the other hand, unlike Bob Park, I am fundamentally a supporter of manned space exploration, and the ISS provides a learning opportunity before more adventurous manned space activity is undertaken. Again my impatience comes to the fore, as progress seems so painfully slow. On reflection, if I were asked to draw up a list of reasons to justify the ISS program, they might be something along the lines of:

- Providing a permanent manned presence in space, and learning how to be there.
- Learning how to build large structures in orbit.

- Learning how to manage large, expensive, complex, and multinational space projects, so that they can be run in an efficient and cost-effective manner.
- Providing inspiration to young people to encourage them to become involved in space engineering and science.

Most of these learning activities will be required in order to take the next steps in leaving Earth and exploring the solar system. I believe that the lessons learned in the ISS program are vital in equipping us for those next steps. However, I am also sure that opponents of manned space activity will still insist that these lessons do not justify the price tag.

Returning to the Moon

The United States has recently declared an intention to return U.S. astronauts to the Moon by the year 2020. This has been spurred by a number of factors, perhaps the main one being a perceived need to regain public enthusiasm for space exploration. Implicit in this statement is a view that a permanent manned presence in low Earth orbit aboard the ISS is not considered sufficiently exciting! I'm sure that motivation has also been provided by the declarations of other nations that have similar intentions. For example, the Chinese space agency have set 2017 as the date for a manned lunar landing, despite the relatively newness of the Chinese manned space program. Another influential feature is the retirement of the Space Shuttle fleet in 2010, forcing major change and new development in the U.S. space program. Although it is important to realize that the U.S. is not the only player in this field, nevertheless I will focus on American plans for a return to the Moon simply because they are sufficiently advanced to give a flavor of how it might be achieved.

The most striking thing about the new NASA plan is that it combines the huge experience gained in the Apollo and Space Shuttle eras of the American space program. The manned vehicle that will replace the shuttle looks very Apollo-like. Initially called the Crew Exploration Vehicle, but now renamed Orion, this spacecraft looks like the Apollo command and service modules (Fig. 10.4). However, it is about three times larger, accommodating four astronauts for the trip to the moon. To launch Orion, a new man-rated launch vehicle is being developed, called Ares 1, using existing components derived from the Space Shuttle and Apollo launch systems. The theory is that this will allow NASA to use tried-and-tested rocket components, and also to benefit from an experienced work force familiar with the manufacture and integration of these components. It is hoped that this will allow a smoother transition to the new operation once the shuttle fleet is retired.

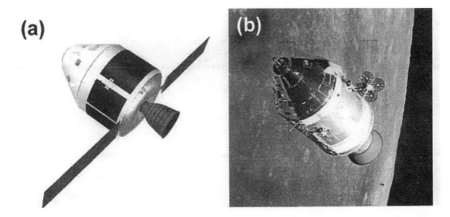

Figure 10.4: The similarity in configuration is apparent when comparing (a) the Orion spacecraft and (b) the Apollo command and service modules, although Orion is about three times larger. (Images courtesy of NASA.)

To land on the Moon, however, requires other equipment, not least of which is a lunar lander spacecraft. NASA proposes developing a separate heavy lift launcher, again composed of Space Shuttle and Apollo components, capable of lifting a payload of the order of 125 metric tonnes into low Earth orbit. This launch vehicle, called Ares 5, will be used to lift the lander spacecraft and a propulsion stage into orbit. It's worth pointing out that although the Ares 1 and 5 launchers are discussed here in the context of the return-to-the-moon program, NASA envisages a wider role for these launch systems involving manned space missions to destinations other than the moon.

So, let's have a look at how all of this new infrastructure to take people back to the Moon fits together, and the way it does is strikingly similar to the Apollo moon landings. The heavy lift Ares 5 launcher blasts off first, taking the unmanned cargo—the lander spacecraft and the propulsion stage—into Earth orbit. Then sometime in the following 30 days the crew takes off in the Orion spacecraft as the payload of the Ares 1 launch vehicle. This new man-rated launch system is much simpler than the Space Shuttle, and as such it is hoped to be more reliable. Once in Earth orbit, the Orion spacecraft will rendezvous and dock with the lander and propulsion stage. After checkout, the propulsion stage rocket engines are fired to boost the whole assembly into a lunar trajectory. Once this maneuver is completed, the propulsion stage is discarded, leaving the Orion spacecraft and the lander docked together for a 3-day cruise to the Moon.

On arrival at the Moon, the main engine of the Orion spacecraft is ignited to take the assembly into a low Moon orbit (Fig. 10.5). The four astronauts

Figure 10.5: The Orion spacecraft docked with the lunar lander in moon orbit. (Image courtesy of NASA.)

then transfer to the lander spacecraft, and undock from Orion, allowing the lander descent engines to fire to take the crew down to the surface. Initially, it is planned that the astronauts will spend 7 days on the surface, while the Orion spacecraft is monitored and controlled robotically in its orbit above. On conclusion of the surface exploration, the crew returns to orbit in the lander ascent module, and then rendezvous and dock with the waiting Orion spacecraft. After transfer of the crew, and the jettisoning of the lander, the main engine of the Orion vehicle is fired to take the astronauts on a trajectory back to Earth. Finally, an atmospheric entry and a parachute descent bring the crew safely home on American soil, probably in California.

Although I personally find the prospect of this shift back to space exploration (as opposed to space applications) exciting, the question of why we should return to the Moon should be addressed. Scientists will always be able to come up with ideas about the scientific exploitation of the unique properties of a lunar base, but do they justify the cost? The price tag for this U.S. initiative is estimated to be about the same as that for the ISS, around $100 billion at least. The opponents of manned spaceflight are hard at work once again bringing criticism to bear on the return-to-the-Moon initiative and claiming how much better value there would be in unmanned, robotic exploration of the Moon.

Another criticism leveled at the new plan is that it has the appearance of a

Majestic Saturn above an ethereal landscape of Titan, as depicted by the imagination of Chesley Bonestell. This is a colour rendition of Figure 10.1 on page 212. Such artist's impressions were an inspiration to me, when I was growing up in the 1950s, before the exploration of the solar system had begun. Image courtesy of Bonestell Space Art.

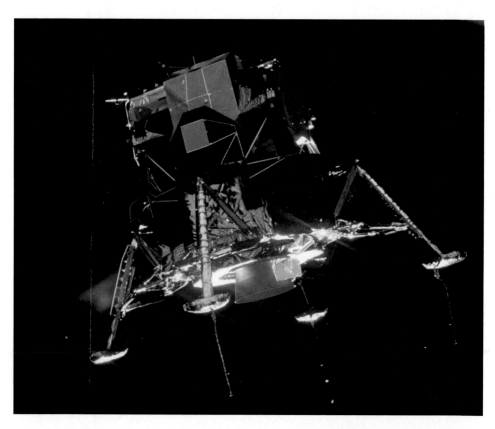

Those magnificent men in their flying machine—the Lunar Module Eagle, with Neil Armstrong and Buzz Aldrin aboard, in lunar orbit prior to their historic descent to the moon's surface in July 1969. The three years of the Apollo moon landings corresponded with my time at university, and I remember it has a time of great optimism and excitement, which was influential in shaping my career ambitions. Image courtesy of NASA.

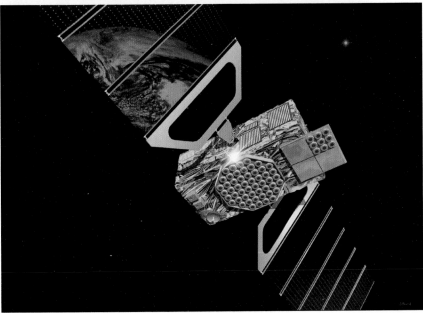

In the decades after Apollo came the boom in space applications. **Communications** and **navigation**, in particular, have had a major impact on people's everyday lives. The upper picture shows Hot Bird 8, a geostationary Earth orbit communication satellite. The spacecraft is shown without its solar array in ground test facility prior to launch. Image courtesy of EADS Astrium. The lower image depicts one of the spacecraft in the Galileo satnav constellation proposed by the European Union. Image courtesy of ESA.

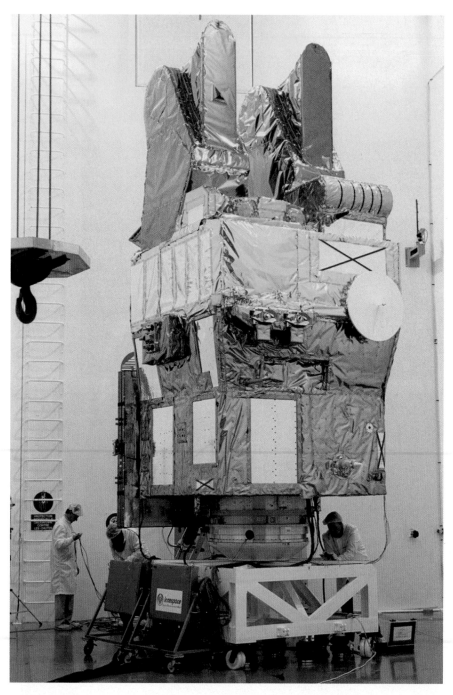

The space applications boom has also included an armada of **Earth observation** satellites in near-polar low Earth orbits. This is a colour reproduction of Figure 9.15 on page 199, showing the SPOT 5 spacecraft prior to launch—see page 154 for details of the spacecraft. The image also shows how the thermal control subsystem dominates the vehicle's appearance—see pages 198 to 206. Image copyright © CNES/Patrick Dumas.

As well as communications, navigation and Earth observation, **science** has of course been a major driver in the development of space technology. The images above, of Saturn and its system of rings and satellites, were taken by Cassini/ Huygens—see page 156 for details of the spacecraft. The upper picture shows a beautiful image of the moon Tethys, which is 1071 km (665 miles) in diameter, against a backdrop of the rings. The lower picture shows Saturn in front of the Sun, with the ring system beautifully backlit. The arrow shows a small smudge, which is the Earth—all of humanity and its affairs confined to 4 pixels on a digital photo! Images courtesy of NASA/JPL—Caltech.

The launch of space observatories, such as the Hubble Space Telescope (HST), has had a major impact on our understanding of the universe in which we live. The image depicts the James Webb Space Telescope (JWST) which is the next generation beyond Hubble. The observatory, which will have a mass around 6,500 kg, is named after the NASA administrator James Webb who was in post during the period 1961 to 1968. The JWST is an extremely ambitious program, involving launching the telescope, in around the year 2013, to the L_2 Langrangian point 1,500,000 km (930,000 miles) from Earth. The main mirror of the telescope is nearly 3 three times larger than that of the HST. The rather strange bed-like structure is a thermal shield about 260 square meters in area. Image courtesy of ESA.

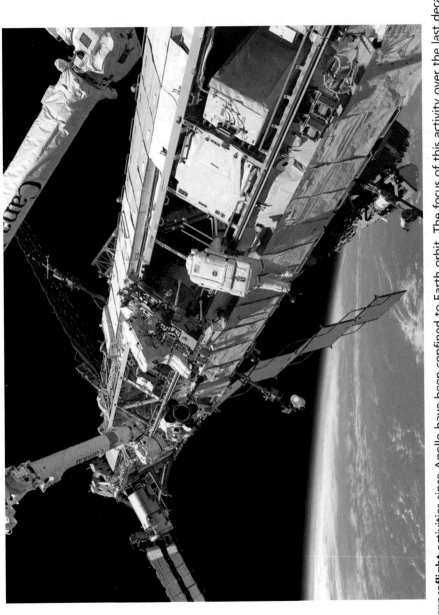

Manned spaceflight activities since Apollo have been confined to Earth orbit. The focus of this activity over the last decade or so has been the construction of the International Space Station (ISS) in a near-circular low Earth orbit with an orbital inclination of 52°, and at an altitude of about 350 km (220 miles). The image is reminiscent of Kubrick's *2001—a Space Odyssey*, with space-walking astronauts working on large space structures. To progress beyond Earth orbit, we need to learn to live and work in this environment. Image courtesy of NASA.

What will **the future** hold? It is always difficult to know, of course, but one space ambition on the agendas of all the major space agencies is a manned landing on the planet Mars within the next thirty years or so. The images depict human exploration of the Martian surface, with astronauts (in the lower picture) visiting the historic site of one of the robotic Viking landings in 1976. Artist's impressions by Pat Rawlings. Images courtesy of NASA.

one-shot moon landing program, similar to Apollo, with no prospect of a lasting legacy, such as the establishment of a longer-term manned presence on the Moon. In answer to this, NASA claims the possibility of building some form of semipermanent lunar base using the new launcher and spacecraft infrastructure. Presumably the cargo lofted by the heavy lift launch vehicle can be modified to take components of such a base to the lunar surface. I think this is perhaps the key to justifying the expenditure: the development of a manned lunar outpost in the long-term for exploration of the Moon and to support manned missions to Mars. In addition to the scientific return that this project would give, the Moon can also serve as a proving ground for a broad range of space operations and processes, including the idea of learning to live off the land—in other words, learning the techniques of self-sufficiency that will be useful in establishing future manned bases in other places in the solar system.

Manned Mission to Mars

The first of these "other places" that the space agencies of the world have their eyes fixedly focused on at the moment is the planet Mars. Compared to other possible landing sites in the solar system, Mars is a relatively hospitable planet, with an atmosphere and a reasonable temperature. However, the emphasis here is on the word *relatively*, as future Martian surface explorers will still require the protection of space suits. The atmosphere is composed of mainly CO_2 (carbon dioxide), and the surface air pressure is less than 1% of that on Earth. Despite the tenuous nature of the atmosphere, winds often whip up dust storms that cover large areas of the Martian surface for several weeks at a time. The approximate average surface temperature is a frigid $-50°C$, and there is also concern that the Martian dust itself may be toxic to humans. And to top it all, due to the fact that Mars's magnetic field is very weak, the surface is pervaded by a flux of solar particle radiation that is attenuated only by the thin atmosphere.

Put in this way, it does make you wonder why people want to invest huge amounts of time, effort, and money to reach Mars! But it is unquestionably the next obvious step in the enterprise of manned exploration of the solar system, with destinations such as Venus and Jupiter ruling themselves out on the basis that they are even more inhospitable. The other main spur for a manned mission to Mars, from the science point of view, is the quest to find evidence of life there. We are not talking about little green men, but more likely the discovery of microbial life. The scientific community is wildly excited about this prospect, since it would tell us something about the occurrence and nature of life in places other than Earth. Again, the anti–manned spaceflight lobby is active in pointing out that this can be done equally well by robotic explorers on the Martian surface, so I guess at the end of the day it is going to be difficult to justify a manned Mars landing on this basis.

This brings us full circle once again to the issue of cost. The cost of a manned landing on Mars is difficult to estimate at this time, but incredible numbers like $1 trillion have been suggested—an order of magnitude increase in spending compared to the ISS or the return-to-the-Moon programs! Clearly, it is difficult to justify this magnitude of expenditure, other than to say things like "it is our destiny." And I think whether we finally go to Mars will hinge on the willingness of the international space-faring community to make this kind of financial commitment.

So how can it be done? Despite the fact that a manned Mars landing may be 30 years away, surprisingly a significant amount of work has been done by space agencies to answer this question. For example, both NASA and ESA (and other space agencies) have developed so-called *reference missions*, to define a Mars landing strategy, and to identify the technologies that will be required to enable the strategy to succeed. Although the reference missions differ in detail, there is nevertheless something of a consensus about the overall approach needed to land people on Mars. Surprisingly, the technology needed for this trip is available now, although some new technologies have been identified that could possibly allow the objective to be achieved at lower cost. The strategy discussed below is a mix of ideas from the various reference missions, but it gives a good idea of how people are thinking about tackling the job.

The basic strategy hinges on the idea of separating the transportation of crew from that of cargo. One day, perhaps 30 years hence, the momentous journey will begin with the unmanned launch of cargo into Earth orbit. At least two such launches will be needed—one to carry an Earth Return Spacecraft, and the other to carry the Surface Cargo Module. Each of these payloads will be of significant mass, on the order of around 150 metric tonnes. After checkout, each spacecraft will be boosted independently out of Earth orbit into a trajectory to take them on their way to Mars. Both of these unmanned elements will use a slow trajectory to Mars, effectively the Hohmann transfer that we talked about in the propulsion section of Chapter 9. The Hohmann transfer to Mars is shown as the slow trajectory in Figure 10.6. The main attribute of this type of transfer, you may recall, is that the amount of rocket fuel required is minimized, reducing overall costs. Nevertheless, to boost each cargo spacecraft out of Earth orbit requires the rocket engines to provide a ΔV (a change in speed) of about 3.6 km/sec (2.2 miles/sec). If a high-performance, but conventional chemical propulsion system is used, this still means that about 80 metric tonnes of the initial 150 metric tonnes will be rocket fuel.

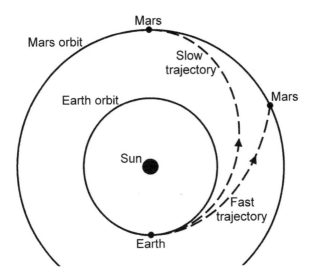

Figure 10.6: Typical transfer orbits to Mars for manned missions. The slow trajectory, the Hohmann transfer, is used for the transit of unmanned cargo. The fast trajectory is used to transfer crew and has a shorter transit time to reduce the effects of microgravity and radiation.

After a trip of 259 days, each of the unmanned spacecraft finally approach Mars on a hyperbolic trajectory (see Chapter 4). To prevent them from just swinging by the planet, their speed needs to be reduced so that they can be captured in an orbit around Mars. The obvious thing to do at this point is to fire rocket engines to nudge each spacecraft into orbit, but this again would cost a significant mass of rocket fuel. To save this fuel mass, some reference missions propose the use of *aerobraking* to achieve Mars orbit. The key to this is the Martian atmosphere. On arrival, each spacecraft dips into the atmosphere and uses the resultant aerodynamic drag to slow down into Mars orbit. Of course, there is a price to pay for this: each spacecraft will require some kind of shield to protect it from frictional heating caused by the rapid passage through the atmosphere. However, calculations show that the mass of this thermal shield is less than the amount of propellant that would be required if the orbit were achieved through firing rockets.

At this point, the paths of the two cargo vehicles diverge. The Earth Return Spacecraft will stay in orbit for a period of years until the astronauts—who at this point are still on Earth—return from their surface exploration. Its job is to take the astronauts home once the mission is completed. On the other hand, the Surface Cargo Module is destined for the Martian surface, where it will wait for the human crew to arrive. As the name implies, this will carry all sorts of equipment that will be useful for the human crew, such as gear for research and exploration, an electrical power plant probably in the form of a nuclear reactor, materials for extending living and laboratory space, surface rovers, and an ascent vehicle to allow the crew to return to Mars orbit at the end of the surface mission. There is also the rather speculative idea of including a fuel production plant to manufacture methane and liquid oxygen from local resources on the surface to fuel the ascent vehicle. The mathematics suggests that an overall saving in mass can be achieved by doing this, but it may not do much for the peace of mind of the crew! So far, so good, but the manned mission has yet to begin!

Once the unmanned spacecraft are in their proper places on and around Mars, and have been checked out to make sure they are in working order, the manned part of the mission can begin. However, to await a suitable planetary alignment, this part of the project will not begin until about 3 years after the launch of the unmanned elements. The Crew Transit and Habitation Module, with a mass of about 150 metric tonnes, will be lofted to low Earth orbit without a crew by a heavy lift launch vehicle. This will be followed within a few days by the crew, probably in an Orion capsule. Once in orbit, the Orion spacecraft will rendezvous with the Crew Transit and Habitation Module to allow the crew to transfer ready for departure. To reduce the travel time to Mars, the Crew Transit and Habitation Module will be boosted into a fast trajectory, as shown in Figure 10.6, at the expense of increased ΔV and fuel mass. This price is considered to be worth paying, in order to shorten the voyage to about 130 to 150 days, so that the harmful effects of microgravity and radiation exposure on the crew can be reduced.

Another way of decreasing the physical effects of weightlessness is to use *artificial gravity* onboard the vehicle. This is the idea of using rotation to produce the sensation of weight. You may have seen films of astronauts being tested for the effects of high launch accelerations by sitting in a big centrifuge. As the speed of rotation of the centrifuge increases, the unfortunate occupants sense a steady increase in their effective weight. In the early days of spaceflight, when astronauts were made of "The Right Stuff" (to quote Tom Wolfe's title of his book about the early astronauts), these machines were used to subject astronauts to levels of acceleration of 8g (and beyond), when the subject effectively weighs eight times their normal weight. Engineers have considered the idea of installing centrifuge-type devices on manned spacecraft destined for the planets so as to combat the physical effects of long-term weightlessness. Another variant is to design the spacecraft so that it consists of two modules attached by a tether system. The two parts are then set in rotation about each other so that the astronauts in each module experience artificial gravity, the level of which is dependent on the rate of rotation. This technology, however, was considered to be inappropriate for the relatively short voyage to Mars, principally on the grounds of increased complexity, mass, and cost.

On arrival at Mars, the Crew Transit and Habitation Module will aerobrake into orbit and descend to the surface, landing within easy walking distance of the waiting Surface Cargo Module so that the surface exploration mission can begin. Figure 10.7 shows an artist's impressions of Martian surface exploration. Hopefully, one day the artist's brush will be replaced by the actuality of photographic images! When the surface stay is over, the crew returns to Mars orbit to dock with the waiting Earth Return Spacecraft. The ascent vehicle is then jettisoned, before the Earth Return Spacecraft is boosted out of Mars orbit for the cruise home. The final act of the mission is a direct entry into Earth's atmosphere, and a parachute descent to a safe landing.

Even in this brief outline, the equipment list for the Mars mission is more extensive than that proposed for Moon missions. This is because the length of stay on the surface of Mars is necessarily much longer, resulting in the need to establish a semipermanent manned outpost during the first landing. This length of stay is dictated by the physics of the motion of the planets around the Sun. To return to Earth, the crew will have to wait for a particular planetary alignment between Mars and Earth, so that the first stay on Mars will probably be many months in duration.

Figure 10.7: Two astronauts explore the Martian surface in an open rover. Artist's impression by Pat Rawlings. (Image courtesy of NASA.)

Manned Exploration of the Solar System

Looking to the future of manned exploration of the solar system beyond Mars becomes a bit of a crystal ball–gazing exercise. For national space agencies, the far future in terms of space program planning means the years 2030 to 2040. Consequently, these plans include a manned mission to Mars, but frankly nothing much beyond that. So to talk about future missions involves guesswork, most of which will miss the mark. However, there are some issues concerning manned spaceflight that are common to the missions we have discussed so far. For such missions to be adequately justified, and ultimately to succeed, there are four key components:

- A good supporting case. These questions need to be addressed: What's it for? Why go? What are the benefits of doing it? Most space exploration is justified on the basis of scientific goals, but other political factors, such as national prestige, work force utilization, and economic benefit through spin-off need to be recognized as valid driving influences.
- An effective team with a common vision of the objectives and how the program can achieve them. With the kind of huge-scale space engineering projects that we have been discussing, such a team will most likely consist of a large number of different nations to share technical responsibilities and program costs.
- The means to go—in other words the technology required to achieve the objective, such as an appropriate launcher capability and manned spaceflight infrastructure.
- An appropriate source of funding. Such missions generally require a huge source of funds sufficient to finance the program, and consequently complete financial planning for the program is required to ensure success. All too often in the past, space projects have suffered from a "stop-go" mentality in terms of funding, governed by political short-termism.

Given the projected cost of a manned landing on Mars, for example, it seems inevitable that deep space missions in the longer-term will continue to be government-sponsored (as opposed to privately funded), and mostly motivated by scientific objectives. Also, we have seen that a good proportion of this cost is driven by the problem of access to Earth orbit—the cost of launch. Currently, this is estimated to be somewhere between about $2000 and $5000 per kilogram launched into low Earth orbit. The other major technical challenge is the development of new space propulsion systems for

Table 10.1: A wish-list of proposed manned space exploration in the 21st century

Year	Mission
2020	Return to the Moon
2030	Manned landing on a near-Earth object*
2035	Permanent lunar base
2040	Manned landing on Mars
2040	Introduction of a single-stage-to-orbit man-rated launch vehicle
2070	Manned landings on the moons of Jupiter (Europa)
2090	Permanent Martian base
2090	Manned landings on the moons of Saturn (Enceladus)

* A near-Earth object is a small body (such as an asteroid or a comet) which has an orbit that comes close to, or crosses the orbit of Earth. Such objects pose an impact threat to Earth (see Chapter 11).

use when the manned spacecraft are beyond Earth orbit. We will discuss these aspects in the next section.

However, let's return briefly to our crystal ball–gazing activity, and ask what kind of missions might be achieved in the 21st century. Table 10.1 lists the anticipated missions over this time period, although it is probably better to call it a wish list, given the shortcomings of the process of so-called prediction in the space business.

Looking at manned missions beyond Mars, exploration of the icy moons of Jupiter would seem to be the next most obvious step. The four major moons of Jupiter—Io, Europa, Ganymede, and Callisto—were discovered when Galileo turned the first telescope in Jupiter's direction about 400 years ago. With Jupiter being five times more distant from the Sun than Earth, the level of solar illumination and heating is around 25 times less (the inverse square law again!) than at the Earth. Generally Jupiter's moons are rather cold, inhospitable places, with the surface temperature of Europa, for example, being around −160°C. However, as unlikely as it might seem, Europa has been identified as a place in the solar system where life may have evolved, and scientists are enthusiastic about the idea of sending robotic and ultimately manned missions there.

The story of potential life on Europa is an intriguing one, which began with the entry into Jupiter orbit of the unmanned Galileo spacecraft in September 1995. Soon afterward, the spacecraft returned images of the icy surface of Europa, such as that shown in Figure 10.8. At first glance, the image looks rather uninteresting. But a more careful examination shows that the icy surface of Europa has fragmented at some time in the past into

Figure 10.8: An image of the icy surface of Europa, taken by the Galileo spacecraft. (Image courtesy of NASA/Jet Propulsion Laboratory [JPL]—Caltech.)

icebergs floating on a liquid ocean, which appear to have drifted off-shore before the ocean refroze. The implication of this is the belief that beneath the icy crust of the moon there is an ocean of liquid water! It is thought that the water remains liquid because it is heated from below by hot volcanic vents on the seabed. As Europa orbits Jupiter, it is subject to tidal forces that squash and stretch the moon, and it is thought that this drives the volcanic activity. How does this lead us to believe there is life there? Well, similar volcanic vents have been found deep in Earth's ocean trenches. These vents are so deep, in fact, that there is no solar energy to sustain life, but marine biologists have found life there, nurtured and maintained by the heat and energy from the volcanism. Scientists believe that a similar process may be happening in Europa's ocean, and the exciting thing is that the life there may have evolved beyond the purely microbial. Arthur C. Clarke picked up on this fascinating idea some years ago in his *Odyssey* series of novels in which he crafts a fine drama around this speculation about extraterrestrial life.

The Cost of Access to Orbit

As we have seen, every kilogram of people, hardware, and propellant that is lifted into Earth orbit costs several thousands of dollars. It is mainly this huge cost of access to Earth orbit that results in the astronomical sums of money needed to fund manned exploration missions. Finding a solution to this fundamental obstacle would result in opening up the new space frontier. However, it is a hard nut to crack.

The ideal solution would be the development of a "Beam me up, Scotty" machine, similar to that used in the *Star Trek* movies. Scientists are giving serious thought to this, and some of them believe that such a transportation device could be operational by the turn of the 22nd century. Current progress on this is slow, but physicists are seriously engaged in developing experiments to demonstrate the transportation of single atoms—I guess you have to start somewhere.

Man-Rated Launchers

In the absence of transporter beam technology, we are back to rocket systems as a means to reach orbit. One way to approach the problem is to divide it into two strands—one dealing with the launch of people, and the other with lifting large amounts of cargo into orbit. If we look first at the issue of launching crew into orbit, there are new initiatives underway (as we have seen) that are at opposite ends of the spectrum in terms of complexity. On the one hand, with the retirement of the Space Shuttle, there is the development of the new manned Orion spacecraft, accompanied by the man-rated Ares 1 launch vehicle. The philosophy here is one of trying to decrease cost, and increase reliability by going back to a simple launch system with a viable escape system for the crew. Fundamentally, the Space Shuttle is a complex machine, and NASA has found it hard work and very costly maintaining an acceptable reliability of 99%, if indeed this can be considered acceptable for a man-rated launcher. On the other hand, there is the approach of developing the complex single-stage-to-orbit (SSTO) launch system that we discussed in Chapter 5. There is no doubt that the cost of access to orbit for crew (and indeed small unmanned payloads) would be considerably reduced, as the goal of such a program is to develop a system that is totally reusable with the operating characteristics of an airplane. Despite the severe technical challenges this poses, it does not seem unreasonable that such a vehicle will be developed through civil and military research programs within, say, the next 30 years. As such, it could be easily integrated into the manned exploration programs around 2040 and beyond.

Unmanned Cargo Launchers

The problem of low-cost access to orbit of large unmanned payloads is more difficult, however. I suppose it is possible that a large cargo version of the SSTO vehicle may be developed, but given the technical difficulties of developing such a vehicle anyway, it seems unlikely that it will ever be able to carry a large payload mass—say, in excess of 10 metric tonnes—into orbit. To solve this, one option is to take the brute force approach of developing large, unmanned cargo carriers, using tried and tested rocket components. The aim would be to develop a reliable heavy-lift workhorse, able to loft payloads of the order of 150 to 200 metric tonnes into low Earth orbit to satisfy the requirements for future manned exploration programs. It is difficult to predict the likely cost per kilogram of payload into orbit, but it would seem reasonable to have a design target of halving the launch costs.

Space Elevators

Another, perhaps longer-term, solution to the problem of reducing the cost of access to orbit is the space elevator. This is the idea of having a cable stretching from Earth's surface to a point beyond geostationary orbit (Fig. 10.9). The cable is anchored to Earth's surface at the equator, and the length of the cable is such that the forces due to Earth's rotation, which tend to fling the cable outward away from Earth, are exactly countered by the weight of the cable, tending to cause it to fall toward Earth. In this way, there is always tension in the cable, so that it will stay erect above the anchor point. Some kind of elevator vehicle can then climb the cable like a beanstalk to carry payloads to orbit. Once the elevator reaches geostationary height— around 36,800 km (22,200 miles)—the payload becomes weightless, so that launching it into orbit is just a matter of gently nudging it out of the elevator. Such a structure has the potential to deliver crew and cargo to orbit at a fraction of the cost incurred by using rocket-powered launchers.

Although this arrangement sounds unlikely, the theory is sound and such a structure is possible in principle. However, there are all sorts of practical difficulties that put its construction beyond our reach at present. The main one is that we currently do not have a material that is light enough and strong enough to withstand the tension in the cable, which reaches a maximum at geostationary orbit altitude. For example, the requirement

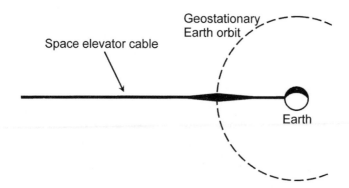

Figure 10.9: A diagram of the space elevator (not to scale). The cable is likely to be tapered, being fattest at geostationary height where the tension is greatest.

exceeds the ability of steel to sustain such a structure by a factor of the order of 100. Materials scientists are working to meet this challenge, but unfortunately it looks like it will be some time before the space elevator solves the problem of low-cost access to Earth orbit.

Space Propulsion

As we have seen, the primary rocket engines used on launchers need to be big to produce enough thrust to lift the weight of the vehicle. Generally speaking, launching things vertically from Earth's surface tends to be a brute force affair, with thrust levels measured in multiples of MegaNewtons, and flight times confined to a few minutes. However, once in orbit around Earth, or in interplanetary cruise, we have a bit more flexibility, in that vehicle weight is not so much a factor so that thrust levels can be smaller and burn times longer.

To date, the vast majority of unmanned exploration of the solar system has been achieved using chemical space propulsion, and doing a simple calculation it can be shown that the best ΔV (change in speed) we can get from an unstaged chemical propulsion system is around 10 km/sec (6.2 miles/sec). This is on the basis of a spacecraft that consists of only the rocket engine, the fuel tanks, and the propellant; it has no payload! Although this is a rather pointless spacecraft, nevertheless it makes the point that currently spacecraft and mission design is significantly constrained by the existing status of propulsion technology. When we look at future missions to the planets, such as landing robotic and human explorers on the surface of Europa, the ΔV requirements are well in excess of 10 km/sec. This is a fundamental obstacle that needs to be overcome by technical developments in space propulsion, and in this section we take a brief look at some of these. Currently, there are two main contenders: nuclear electric propulsion and nuclear thermal rockets.

Nuclear Electric Propulsion (NEP)

Electric propulsion systems have been developed over many years and have already flown on unmanned spacecraft. There is a fundamental difference between chemical and electrical propulsion systems. With a chemical system the energy required to accelerate the propellant out of the rocket nozzle is obtained from burning the fuel/oxidizer combination. The speed we can impart to the vehicle is fundamentally limited by the amount of energy contained in the chemical propellants. However, an electric propulsion system is separately powered, inasmuch as the energy to accelerate the

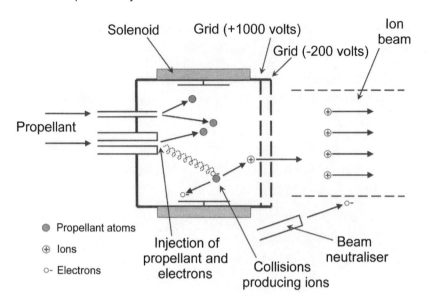

Figure 10.10: A cross-section through the cylindrical ionization chamber of an ion drive.

propellant comes from a separate source, and so in principle is unlimited. As the name implies, this separate source is electricity, and this can be generated using sunlight (solar panels) or nuclear energy.

A common form of electric propulsion, called the *ion engine*, is shown in Figure 10.10, which depicts a simplified cross section through a typical ion engine. The engine is a squat cylindrical chamber, a bit like a paint can, into which the propellant is injected. In the ion engines that are used in small unmanned spacecraft, the diameter of the cylinder is around 10 cm (4 inches) in size, and the usual propellant is an inert gas, such as argon or xenon. As well as the propellant, electrons are also injected into the chamber from a heated hollow tube, called a hollow cathode. Surrounding the cylindrical surface of the chamber are solenoids—basically electromagnets—that produce a magnetic field within the chamber, causing the electrons to corkscrew around the magnetic field lines (see Figure 6.7 in Chapter 6). In this way, the likelihood of collisions between the electrons and the propellant atoms is increased, and these collisions cause ionization of the propellant. In other words, electrons are stripped from the propellant atoms, causing them to acquire a positive electric charge. These positively charged propellant atoms are called ions. (There are a variety of ways of producing the population of ions in the chamber other than the use of a hollow cathode described here.) At the exit of the device, there are metal grids to which a voltage is applied to accelerate the ions out of the device, forming a high-speed ion beam. Finally, to prevent the spacecraft acquiring a huge negative electric charge, electrons are squirted downstream into the ion beam by a beam neutralizer, to allow the electrons to recombine with the propellant ions.

The resulting exhaust velocity of the device is typically on the order of 50 km/sec (30 miles/sec)—about 10 times higher than that of a high-performance chemical system—but the achievable mass flow rate is much smaller. As a consequence, a typical ion engine has a large specific impulse (which is good), but a low thrust level (which is not so good). Recalling the discussion about specific impulse in Chapter 5, this means that, all other things being equal, the ion engine will produce about 10 times more ΔV (speed change) for a given mass of fuel than a chemical system. However, the low thrust means that it will take a long time to do so. Fortunately ion engines can operate for thousands of hours, so the tiny accelerations that they produce can build up large ΔVs, but one has to be patient. It is a 0 to 60 mph in 2 weeks kind of performance! For an ion engine powered by solar panels, with an input of around 1 kW of electrical power, the level of thrust is quite tiny—on the order of 1/20th of a Newton!

To date, this kind of system has been used on unmanned spacecraft for things like orbit control of spacecraft in Earth orbit, or for missions to the Moon and near-Earth asteroids. But the question is, Can the system be scaled up to produce a useful means of propelling large manned spacecraft? The obvious route is to use a nuclear reactor power system to increase the power levels to the hundreds of kW level. Research into the feasibility of such NEP systems has been underway for many years, but such a system is yet to be flown in space. Using some simple calculations, we can scale up the 1 kW system mentioned above to a power input of, say, 500 kW. The configuration of such an ion drive is speculative, but we can envisage this power input supplying maybe five ion engine units each 60 cm in diameter. A quick calculation gives a thrust level of around 15 N, which (if we recall our informal definition of a Newton) is a force equivalent to the weight of about 15 small apples. Although this is more useful than the 1/20th of a Newton we had before, even so it is questionable whether an ion drive can be scaled to provide a propulsion system for manned spacecraft, which tend to be large (of the order of 100 metric tonnes in mass). We will come back to this in our summary later.

Another issue that affects the dynamics of nuclear powered propulsion systems is the mass of the power plant. The nuclear reactor is likely to be of a significant mass, which will add to the already burgeoning mass of a manned spacecraft. This is hard to estimate, but the mass of a 500 kW reactor, for example, might be around 3 metric tonnes.

Nuclear Thermal Rockets (NTRs)

The nuclear thermal rocket is another propulsion technology where the energy is provided by a separate source, once again a nuclear reactor. But the mode of operation is different from that in the NEP system. To operate a NEP device, a reactor is used to produce electricity to power the system. However, in the NTR the heat produced by the nuclear reactor is not converted to electricity. Instead it is used directly to heat the rocket propellant, energizing it to produce thrust. NTRs have been developed and tested over many years, but again the technology has not been flown in space.

The NTR shown in Figure 10.11 is referred to as a *solid-core configuration*, which is the simplest design to construct. In concept, the operation of the rocket is relatively straightforward. Liquid hydrogen propellant is passed through the reactor core, acting to cool the reactor and to heat the hydrogen to temperatures of around 3000°C. This superheated hydrogen is then expanded out of the engine nozzle to produce thrust. This engine can produce a high thrust for relatively long periods of time, up to about an hour, and has a specific impulse of around 1000 seconds. This specific impulse is an improvement over the best chemical propulsion by a factor of two, so that about twice the ΔV is achievable for a given mass of propellant. To give an example of what this means in terms of numbers, let's suppose we use a 200 kN thrust solid-core NTR with a specific impulse of 1000 seconds to propel a spacecraft with an initial mass of 150 metric tons. If we fire the engine for an hour, the burn would result in a ΔV of 6.5 km/sec (4.0 miles/sec), so that we are getting a performance that is really useful for the propulsion of manned exploration missions. The mass breakdown for this example is about 73.5 metric tonnes of fuel, 6.5 metric tonnes of NTR, and 70 metric tonnes of useful payload. Also, by using different designs of NTR, the specific impulse can be further increased up to around 2000 seconds, so in principle the mass of fuel can be reduced further.

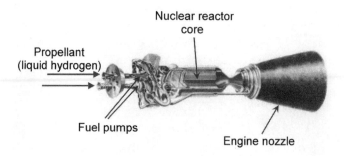

Figure 10.11: A cutaway diagram of a solid-core nuclear thermal rocket. (Backdrop image courtesy of NASA.)

Comparison of Space Propulsion Technologies

To compare the space propulsion technologies we have so far considered, I have taken the example of the Earth departure engine burn to transfer a manned spacecraft onto a trajectory to Mars. Table 10.2 compares the propulsion technologies; we can see that a fixed ΔV of 3.6 km/sec (2.2 miles / sec) and a fixed initial vehicle mass of 150 metric tonnes are assumed in all cases. The chemical engine is equivalent to one Space Shuttle main engine, and the NEP system is assumed to be powered by a 500-kW nuclear reactor. The characteristics of the NTR are those adopted by NASA in its Mars reference mission.

The table shows that in terms of minimizing propellant, the NEP ion drive wins easily. However, the thrust level of the ion drive is so small that the time to execute the rocket burn is ridiculously long, ruling it out as a practical proposition. This issue effectively discounts the use of NEP for manned exploration, where vehicle masses are generally large. The use of NEP is perhaps more appropriate for operating large unmanned interplanetary spacecraft in the 20-metric-tonne class. Such projects have been proposed in the past for missions such as the robotic exploration of Jupiter's system of icy moons, but they have yet to get off the drawing board. The table shows that the NTR is promising in terms of propulsion for manned missions, with not only an appropriate thrust level but also a useful payload mass.

However, there are other issues with operating nuclear powered spacecraft that we have not yet mentioned. The first and most obvious, perhaps, is the issue of the radiation that the reactor produces, and the harmful effects that this has on the crew. We have already seen that the

Table 10.2: A comparison of propulsion technologies, based on Earth departure ΔV for a Mars mission. Note: An initial mass of 150 metric tonnes is taken for all cases, as being representative of a manned spacecraft.

Type of engine	Chemical	NEP	NTR
ΔV (km/s)	3.6	3.6	3.6
Initial mass (metric tonnes)	150	150	150
Specific impulse (sec)	450	5000	960
Exhaust velocity (km/sec)	4.4	49.1	9.4
Thrust (N)	2,000,000	15	200,000
Burnout mass (metric tonnes)	66	139	102
Fuel mass (metric tonnes)	84	11	48
Burn time	3 minutes	393 days	37 minutes

mass of the reactor is significant, but there is also the requirement for additional mass in the form of radiation shielding to protect the astronauts. Finding the best place for the reactor, and the accompanying shielding, relative to the crew's living space will have a major impact on the vehicle design. A significant mass of thermal radiators will also be needed to ensure that the thermal output from the reactor system is adequately dissipated. A final issue, perhaps of a political nature, is that the NEP and NTR systems both carry the label "nuclear." Although it is intended to use these nuclear-powered systems only in space, where any harmful environmental effects are negligible, nevertheless their operation will inevitably attract opposition from the increasingly vocal "Green" lobby. The focus of concern is not so much the space operation, but the requirement to launch the systems into orbit. It is certainly true that a launch failure would scatter radioactive pollution throughout the atmosphere and on the ground.

Although there are other (fairly wacky) ideas to solve the problem of space propulsion, many of these appear to be viable only in the distant future. Unfortunately, none of the more sensible ideas come anywhere near the propulsion systems we see routinely in science-fiction movies like *Star Wars*. The kind of energy source that will allow manned vehicles the size of small airplanes the freedom to launch into space, fly across the galaxy, land on other planets, and then return, simply escapes our ingenuity at the moment. Hollywood space engineering has always been a lot easier than the real thing!

Space Privatization and Space Tourism

Moving away from government-sponsored space programs for a moment, it is interesting to ask what the future holds for space privatization and the expansion of the booming terrestrial tourism industry into the new arena of space.

Space Privatization

In our discussion above, concerning manned space programs, we have painted a picture of huge sums of money being spent over decades of time, where the payoff comes mostly in terms of the advancement of scientific knowledge and possibly political prestige for the participating nations. Put in this way, it is easy to appreciate why it is hard for private industry to get involved, given the normal requirement of a good financial return on investment and within a reasonable time scale. Private industry's need to make a fast buck is a philosophy that does not fit well with the overall enterprise of manned space activity. Perhaps the only way around this roadblock is through the development of space tourism, and this is something we will come back to in a moment.

Despite views to the contrary, there is a good deal of private industry involvement in space activity, and to find out where this is happening, you just need to ask which bits of space turn a profit. The most profitable space application over many years has been that of satellite communications. Over time, intercontinental telephone traffic has increased greatly, and one of the best ways of doing this is through satellites, predominantly in geostationary Earth orbit. And there is a good return on investment to be had by the spacecraft owners. Earth observation is another contender that shows promise in this respect. The idea is a simple one—that of selling satellite images of Earth to users who need the data for a variety of reasons, including searching for Earth resources, weather forecasting, disaster assessment and management, the planning of large civil engineering projects, map making, and even agriculture. A number of private ventures have tried to turn a profit doing this, but this is generally only possible if the costs of the spacecraft and the launching are removed from the equation. Earth observation is not strictly a profitable activity at present, but the potential is there for the future if we can drive down spacecraft and launch costs.

Beyond communications and Earth observation, perhaps the next most promising commercial candidate is satellite navigation (satnav), which we discussed briefly in Chapter 1. We are all familiar with satnav systems these days—for example in cars—that operate using the American military Navstar GPS system. However, in around 2012 a civil system will be launched by the European Union called Galileo, and the intention is to charge users for access to the system. There are some issues here that muddy the waters regarding Galileo's profitability, such as convincing people to pay to use it when there is a perfectly acceptable and free alternate system in the form of GPS. There are some fairly long-standing political and financial issues with Galileo that people are grappling with at the moment, but if these can be resolved, there is great potential for turning a good profit.

Finally, although the list is perhaps not exhaustive, the provision of launch services is the last obvious example of commercial space enterprise. There are a number of commercial launch companies in the U.S., Europe, and Russia that market and sell launch opportunities to governments and private customers. Arianespace, which operates the Ariane family of launch vehicles, became the first such commercial space transportation company in 1980, and continues to be one of the largest providers of rides to orbit.

So, currently there is good evidence of commercial activity in unmanned

Figure 10.12: The X-prize winning SpaceShipOne on its descent after reaching in excess of a 100-km altitude in 2004. (Copyright © 2004 Mojave Aerospace Ventures LLC. Photograph by Scaled Composites. SpaceShipOne is a Paul G. Allen project.)

space applications, but what of the privatization of human spaceflight, given the issues of cost, time scales, and return on investment? A first step was taken in 2004, when a piloted vehicle called SpaceShipOne (Fig. 10.12) won the so-called X-prize of $10 million, which was set up to stimulate private investment in the development of manned spaceflight technologies. To win the prize, SpaceShipOne had to demonstrate the first human spaceflight to an altitude in excess of 100 km (62 miles) in a privately developed and operated vehicle. Although this is a considerable achievement, one very important point should not be missed here: the vehicle reached orbital height but did not achieve orbital speed. As we discussed in Chapter 2, to enter orbit at a 100-km altitude requires traveling horizontally at about 8 km/sec (5 miles/sec), and in terms of technical achievement it is the attainment of orbital speed that is the difficult part. Acquiring this speed demands the input of huge amounts of rocket-powered energy, which makes conventional launch operations so risky and expensive. So in a way you could say that the X-prize competition missed the point. However, as a consequence of this first privately funded manned spaceflight, a number of commercial companies have jumped on the bandwagon, and are proposing to build a number of such spacecraft to take people to orbital height. The motivation for this investment is tourism, and in the first instance, seats will be sold to passengers at around $200,000 each. For this fee, paying customers will experience the view of Earth from 100-km altitude, and a period of a few minutes of weightlessness at the top of the trajectory when the vehicle is in unpowered free-fall. So here, finally, we begin to see a motivation for private investment in manned spaceflight—that of space tourism.

Space Tourism

The first fare-paying space tourist was Dennis Tito, who visited the International Space Station for 7 days in 2001. Since then, a handful of others have followed him to enjoy the experience of life in orbit. However, a ticket costs about $20 million. For Tito, that was about $5000 per kilogram to lift him into orbit, and then a room rate of about $2.8 million per day! Clearly the opportunity to take a holiday in space is not available to most people because of the cost of access to orbit. The suborbital flights mentioned above, with a ticket price of around $200,000, is a way of bypassing this obstacle, but nevertheless cost is still the fundamental barrier to the exploitation of space by the tourism industry.

Inevitably, the key to opening up space to the tourist market is access to Earth orbit that is cheap, reliable, and safe. Once again we have gone full circle, and arrived back at the need for the development of a single-stage-to-orbit man-rated launch vehicle with the reliability and operational characteristics of a civil airliner to open up this new commercial opportunity. As we mentioned in Chapter 5, this is a major technical challenge, requiring considerable investment probably on the back of a military research and development program. If this is how it happens, then it will not be the first time that commercial enterprise has benefited from the fruits of military research. In terms of time scales, I stuck my neck out earlier in this chapter to suggest that such an SSTO launcher may be operational within 30 years or so, so again patience is the order of the day!

This is not the end of the story, however. Once the problem of orbital access is solved, there is still a need to develop space infrastructure in orbit and beyond, to take passengers to their intended holiday destinations. For example, I have always thought how great it would be to take a weekend break to see the rings of Saturn, or to spend a week on an elevated lunar monorail taking in the splendors of the lunar landscape—a bit like a lunar equivalent of a Canadian Rockies scenic railway ride but without the weather (and of course stopping off for a quick snack at the Tranquillity Base

McDonalds along the way). The principal challenge here is to develop a safe and reliable means of space propulsion that will reduce the travel times to the distant planets from our current expectation of years to a few days—a tall order indeed.

No doubt the development of space tourism will be incremental, starting with Earth orbit hotels, and then progressing to the Moon and beyond as space propulsion technology develops. However, it is clear that real space tourism—people vacationing throughout the solar system for a price that the majority can afford—is going to take a long time, and unfortunately I have no expectation that trips to Saturn's rings will be available within my lifetime!

In the next and final chapter, we take a brief look at a couple of issues that relate to the necessity for us to be fully involved in the business of safe and inexpensive access to space.

Space: The Final Frontier?

Deep Impact

ARE we "cleverer" than the dinosaurs? One day, about 65 million years ago, a fireball streaked across the sky above what is now Central America, and impacted the ground in the region of the Yucatan peninsular. The object, traveling at huge speed, was an asteroid about 10 km (6 miles) across, and the enormous energy released by the impact produced global devastation and played havoc with Earth's climate. As a consequence of this meeting of the celestial with the terrestrial, many scientists believe that the 160-million-year reign of the dinosaurs was brought to an end. Could this happen again, with people this time being the victims of potential extinction?

The answer, disquietingly, is yes. It will happen again. But impacts of the magnitude of the Yucatan event fortunately do not happen often—about once every 100 million years. So it is probably something we do not need to worry about for a long time. However, as we develop the technology to look out into space, we are beginning to realize that there are a large number of objects in orbit around the Sun that potentially could collide with Earth. These objects are called *near-Earth objects* (NEOs). A NEO is an asteroid or comet in an orbit that crosses or comes close to Earth's orbit around the Sun, and so represents an impact threat. There is intense activity at present to detect and catalogue the NEO population, and so far we have estimated that there are about 1000 objects on the order of 1 km (0.6 miles) in diameter or bigger. The detection of smaller objects becomes more difficult, so it is not known how many smaller objects there are, but we do know that the number of objects increases as the size decreases. Current estimates of the number of objects bigger than 100 m (330 feet) is about 100,000. As a consequence, we are never quite sure when one of these objects will be discovered to be on a collision course with Earth.

The last significant impact event took place in 1908 at Tunguska, Siberia, and this object was estimated to be about 50 m (165 feet) across. Fortunately, the impact site was uninhabited, but the explosion flattened about 2000 square kilometers (770 square miles) of forest. If the object had landed in

G. Swinerd, *How Spacecraft Fly: Spaceflight Without Formulae,*
DOI: 10.1007/978-0-387-76572-3_11, © Praxis Publishing, Ltd. 2008

central London, for example, the area of devastation would correspond approximately to everything within the M25 orbital motorway. An impactor of this size can be expected about once every few hundred years. Consequently, a NEO impact is a fairly rare event, but nevertheless there is a probability we will have to face one in the not-too-distant future! National governments, charged with the responsibility of looking after their citizens, are now at least considering this type of event as a natural disaster, alongside things like earthquakes and hurricanes. Hollywood has also done its bit to raise awareness with films such as *Armageddon* and *Deep Impact*, which with the aid of computer-generated images give a graphic depiction of some of the devastating impact-generated effects. For land impacts, these include the effects of blast, heat, and ejecta from the impact site, and the generation of seismic disturbances. However, since the majority of the Earth's surface is water, it is more likely that such an object will fall into the ocean. For this type of event, the main impact-generated effect is a large tsumani wave, which propagates at high speed across the ocean. Tsumanis are very effective at transporting the energy of the impact to distant shores, bringing devastation to coastal cities where most of the world's population is concentrated.

So what can be done? Unlike the dinosaurs, we are at least in a position to see something coming, and to do something to deflect its path to avoid a collision. The technology needed to do this is available now, but the usual roadblock is funding. Budgets available for NEO detection surveys are relatively small, and agency budgets for the development of spacecraft that could be used to deflect a threatening NEO are similarly inadequate. If a 500-m object suddenly came over the horizon today on an impact trajectory, threatening devastation on a continental scale, the funding situation would dramatically change. But would there be enough time to develop and test the require space hardware to be assured of success? Can we afford to take the chance? I would suggest not.

There are a number of ways of deflecting the path of a threatening NEO. They all have one feature in common: they depend on there being sufficient warning of an impact, so that the missions can be launched several years in advance. This is because, generally, the methods are only able to produce small changes in the trajectory of the NEO. Such a small change is able to deflect the object successfully if it is done a long time in advance, as the effects of a small change build up over time to avert disaster. However, if the time to impact is short, then much larger changes are required, which are effectively beyond our current capabilities. This is why surveys designed to detect threatening NEOs well in advance are so important. The following list of deflection techniques is not exhaustive, but it does give a flavour of some of the ideas that are being proposed.

- **The use of nuclear weapons:** This is invariably the Hollywood solution, as it makes for good cinema! The strategy is to launch one, or a number of nuclear warheads against the incoming object to blow it off course. The problem here is that no one really knows how the nuclear blast will affect the object's motion, and tests need to be done to find out. In the vacuum of space, the detonation produces a blast wave that is very much less powerful than it would be in Earth's atmosphere, so that the effect of the explosion on the asteroid's motion may be inadequate. However, there is another deflection mechanism that may be effective if the nuclear weapon is detonated in close proximity to the asteroid's surface. Then the surface layer may possibly be heated sufficiently to vaporize the asteroid and blast it at high speed into space. On the basis of Newton's laws (Chapter 1), this ejection of material would produce a thrust on the asteroid, a bit like a rocket engine, which may be sufficient to deflect the object's path from a collision course. There are many unanswered questions about this technique, which emphasizes the need for flight tests to ascertain how nuclear explosions influence the motion of a NEO in its orbit around the Sun. The other issue with the use of nuclear detonations is the risk that the NEO may be fragmented, resulting in a cloud of smaller but still potentially lethal impactors on their way to Earth. In this case, the situation may have been worsened, with a multitude of Tunguska-type impacts bringing worldwide devastation.
- **The use of an impactor:** This is an intuitive idea—crashing an impactor spacecraft onto the surface of the object to deflect its trajectory, like a billiard ball changing its path after the impact from another ball. But the thing to note about the billiard balls is that a significant change is produced because the two balls are of the same size. Obviously, we are unable to launch an impactor spacecraft with the same mass as, say, a 200-m asteroid, so the amount of deflection is tiny. However, if the deflection is done sufficiently far in advance, then a small change will be adequate to avoid a collision with the Earth.
- **The use of a gravity tractor:** This is a rather less intuitive idea, but one of the most effective ways to achieve a controlled deflection of a threatening NEO such as an asteroid, without needing to know anything about its physical characteristics, such as the nature of its surface, or its state of rotation. The idea is a relatively recent one, being proposed in 2005 by Ed Lu and Stan Love of the National Aeronautics and Space Administration's (NASA) Johnson Space Center. The gravity tractor is an unmanned spacecraft that is launched to rendezvous with the asteroid. On arrival, the tractor

positions itself a small distance from the object, and then uses small rockets to hold this position above the surface, as shown in Figure 11.1. As the spacecraft hovers above the surface, a force is exerted on the asteroid equal to the mutual gravitation force between them. Effectively, the tractor is using gravity as an invisible tether with which to tow the asteroid off it collision course with Earth.

To get an idea of the numbers, let's suppose the asteroid is 200 m across. Then a quick calculation gives its mass as about 10,000,000 metric tonnes. If the tractor spacecraft's mass is about 5 metric tonnes and it is stationed 50 m above the asteroid's surface, then it will require a thrust of only about 0.2 Newtons to remain stationary above the asteroid. This small force is equal to the mutual gravitational force between the tractor and the asteroid, and it is this force that produces the acceleration to shift the asteroid from its collision course. The application of such a small force to such a huge mass produces a tiny acceleration. However, the saving grace is that this tiny acceleration can be applied for a long period of time to allow a useful change in speed of the asteroid to build up. If the mission is performed a sufficiently long time in advance of the predicted Earth impact, then the required change in the asteroid's orbit to avert disaster is similarly small, and easily accommodated by the technique. The operation to successfully divert the asteroid in this case can be achieved in about 10 days, but much longer periods can be realized if required.

Figure 11.1: An artist's impression of a gravity tractor spacecraft on-station above an asteroid on collision course with Earth. (Image courtesy of Dan Durda, B612 Foundation/FIAAA.)

In terms of the spacecraft, the propulsion requirements can be achieved by two ion drives, each with a thrust of 0.1 N. The electrical power required for these would be about 4 kW, and the fuel mass used in this case is approximately 4 kg.

As well as the technical issues discussed above, there is also a political dimension to the issue of NEO deflection. Once a potentially threatening NEO has been identified, the first task to be undertaken is to determine its orbit around the Sun so that the likelihood of an impact can be estimated. If the object is indeed on a collision course, the site of the impact on the Earth's surface can be estimated, although if the impact is some years away there are going to be significant errors in this process. However, let's suppose that such a process has led to a likelihood that the object will fall somewhere on the North American continent. In this situation, the United States would be eager to launch a deflection mission as soon as possible to avoid the devastating consequences of such an impact on its territory. However, once the spacecraft has reached the object and the deflection process is underway, say, with a gravity tractor, then the resulting changes in the NEO's orbit will progressively change the location of the impact site. In this situation, should the impact site be moved out over the Pacific Ocean where significant tsunami damage to the west coast of the U.S. can result? Or should it be moved east toward the continents of Europe and Africa? Of course the objective is to move the impact site so that the object misses Earth entirely, but what if the gravity tractor spacecraft fails before its mission is completed? In such a time of crisis, the international community may be called upon to make some monumental decisions that will affect the lives of millions of people. But as yet there is no agreed international mechanism through which such decisions can be made.

As we have seen, the probability of an asteroid impact with Earth in the next few decades is small, but nevertheless if one were to occur the consequences are horrible to contemplate. The prospect of such an event provides a good case for developing our space-faring capabilities. I think it is time that we grasped the nettle and showed that we are indeed "cleverer" than the dinosaurs!

Leaving Home

One day, about 5 billion years from now, the last perfect day will dawn. It is at about this time that scientists estimate the nuclear fuel of the Sun will begin to run out. As we discussed in Chapter 6, the Sun is powered by a nuclear fusion reaction at its core, with hydrogen atoms being fused together to form heavier atoms, and in the process producing the energy that has made the

Sun shine steadily for the last 5 billion years or so. The Sun has shone stably for all that time because of the balance between the huge amounts of energy being generated at its center, tending to blow it apart, and the force of gravity tending to hold it together. On this last perfect day, 5 billion years in the future, the nuclear fuel at the Sun's center will be just about depleted, and the stable balance between energy generation and gravity will be disturbed. The consequences for the Sun will be dramatic as far as the inhabitants of Earth (or indeed any other planet in the solar system) are concerned. I'm not sure who those inhabitants will be; people have been around on Earth for only about a million years, and it seems strange to think of them still being here 5 billion years in the future. Perhaps some other species, directly descended from humans, will exist then, but that's a different story.

What will happen to the Sun when its hydrogen fuel begins to run out? According to our best theories, it will evolve into a red giant star, expanding to a sphere about the same size as Earth's current orbit. In this process the Sun will lose a significant amount of mass, so Earth's orbit radius is predicted to increase to about one and a half times its current radius. Thus Earth will probably escape being engulfed by the Sun. But the surface environment on Earth will be transformed into a blazing desert, with all the oceans' water having boiled away. Put simply, Earth will no longer be able to support life, other than perhaps microbial life buried deep within Earth's crust.

This rather bleak picture of the Sun in its death throes tells us that ultimately people will have to leave Earth. This notion of our successors having to leave home in the distant future has perhaps become a bit of a cliché in contemporary science-fiction literature. Some people believe that other factors, such as climate change, may be more important in forcing the evacuation much sooner. However, the bottom line is still the same: to ensure our ultimate survival as a species, we have to learn how to live and work in space. More importantly, we need to develop and master the technologies required to travel across the cosmos, that is, to transform Hollywood space engineering into reality! How are we going to do this? The technical challenge is huge, simply because the universe is huge. This is expressed rather eloquently by Douglas Adams in his book, *The Hitchhiker's Guide to the Galaxy:* "Space is big. You just won't believe how vastly, hugely, mind-bogglingly big it is." To describe the sorts of distances we are talking about, the nearest star (apart from the Sun) is about four light years away. This is the distance that light travels in 4 years at the enormous speed of 300,000 km/sec (186,000 miles/sec). A quick calculation gives this as about 38,000,000,000,000 km (23,000,000,000,000 miles), a distance probably too large for our minds to comprehend. Using our current spacecraft technology, we know that it takes many years to reach Pluto, the most distant outpost of our solar system. However, the *nearest* star outside the solar system is about 6500 times more distant. This demonstrates the magnitude of the challenge, without even addressing the prospect of traveling across our home galaxy, the Milky Way, which is estimated to be about 100,000 light years across!

How are we to achieve travel across such vast distances? Well, going very fast is obviously a good idea, but our currently accepted laws of physics, due to Einstein, set a speed limit equal to the speed of light (see Chapter 1). Although I think that Einstein might not be the last word in our understanding of the universe, nevertheless in this discussion I will stay within the boundaries of his theories, and accept the light speed limit. To cross huge distances, it would be good to be able to travel at a speed that is a good percentage of light speed, or alternatively find clever loopholes in the laws of physics that will allow us to effectively travel faster than light speed without strictly violating the speed limit. Although the latter sounds like a bit of a contradiction, nevertheless we will see that there are some interesting ideas along these lines (see Exotic Systems, below). Let's take a brief look at some of these ideas for achieving interstellar travel, starting with some less exotic but still futuristic rocket systems.

Rocket Systems

Many of the ideas for achieving what might be called slow interstellar travel are based on using the principle of a rocket, along the lines of a device like the Space Shuttle main engine that uses Newton's third law of motion to operate: for every action there is an equal and opposite reaction (see Chapter 1). Also, when I say "slow," I mean spaceships with a maximum speed of, say, 10% of the speed of light—around 30,000 km/sec. Although this seems fast by the standards we adopt in our everyday lives, it does mean that such a vehicle would still take about 40 years to reach the nearest star outside the solar system. To travel to stars in the local neighborhood, say, 100 light years distant, it is going to take several human generations to get there!

Nuclear Impulse Engines

The first such idea is that of the nuclear impulse engine, which adopts the unlikely notion of the detonation of numerous atomic bombs to accelerate the vehicle. The concept was first proposed in the 1940s, and subsequently fleshed out into a design concept called *Project Orion* in the 1960s. Intuitively the idea is an easy one to understand. Small nuclear bombs with a yield of about 10 metric kilotonnes of TNT (this is about half the size of the atomic bomb that destroyed Hiroshima in August 1945) are detonated behind a rigid pusher plate, which in turn is connected to the starship by a system of springs and shock absorbers. This form of propulsion is unusual in that it has both a high specific impulse (exhaust velocity) and a high thrust level. The impulse of each explosion is transferred to the starship, to accelerate it to an anticipated speed on the order of 10% of light speed. Clearly, for a manned starship, there are issues concerning the protection of the crew from high accelerations, blast effects, and nuclear radiation. However, the severity of these problems are reduced for large vehicles—on the order of 1000 metric tonnes or more—for which the pusher plate can be scaled up to be several meters thickness of steel, and so provide adequate shelter for the crew.

Fusion-Powered Rockets

These rockets have great potential for powering interstellar travel, but since a controlled nuclear fusion reaction has not yet been achieved in terrestrial laboratories, such propulsion technology must for now remain a promising prospect for the future. Recall from Chapter 6 that the Sun is powered by nuclear fusion, where atoms of hydrogen are fused together to form heavier atoms while releasing energy in the process. Nuclear power stations here on Earth generate our electricity needs, but they use a different kind of nuclear reaction—*nuclear fission,* in which heavy elements such as uranium are split to form lighter atoms, which also produces nuclear energy. Our current technology allows us to control the nuclear fission reaction, but taming nuclear fusion to produce a controlled reaction for the purpose terrestrial power generation is something that has escaped our ingenuity so far. Solving this problem would meet the world's energy needs, as there is an inexhaustible supply of fusion fuel, in the form of hydrogen and similar light elements, in the world's oceans. Needless to say, a huge research effort has been expended to try to crack this problem, but unfortunately we have managed to achieve only an uncontrolled nuclear fusion reaction in the hydrogen bomb for destructive purposes. Controlled fusion currently eludes us, but scientists working in the field appear confident that the difficulties can be resolved in two or three decades.

The reason why controlled fusion is so difficult to achieve is the high temperature at which the reaction takes place. To sustain a fusion reaction, the fuel needs to be at very high temperature—millions of degrees Celsius. We have to go some way to emulate the conditions found at the center of the Sun, where the fusion reaction occurs naturally. The fuel is confined within the reactor as a hot gas made up of charged particles, and this is referred to as a *plasma.* Such an extremely hot plasma must be confined in the reactor in such a way as to prevent contact with the reactor walls, which is usually done in terrestrial laboratories by containing it within a magnetic field. However, this confinement within a "magnetic bottle" is the stumbling block; it is difficult to produce a method of magnetic confinement that is stable. Even if this problem were to be solved for terrestrial reactors, for space applications it is thought that magnetic confinement is not the best solution because the mass of such a system would be prohibitive. Alternative methods of sustaining a fusion reaction for a rocket system have been investigated, such as igniting small pellets of nuclear fuel (just a few millimeters across) using electron beams or lasers.

However, whatever method finally proves successful, the point is that a fusion-powered rocket can produce an extremely high temperature plasma, and this can be channeled through a magnetic nozzle to produce thrust. The performance of such a rocket is unknown, but exhaust velocities of the order of 10,000 km/sec may be feasible! Taking this value, a quick calculation shows that a fusion-powered starship could be accelerated to one-tenth the speed of light if 95% of its initial mass is propellant.

Antimatter Rockets

Readers may be familiar with the word *antimatter* from watching too much *Star Trek* on TV. Indeed, in contemporary science fiction, antimatter seems to be the ubiquitous source of energy that solves all future problems concerned with interstellar travel, and in fact there is an element of truth in this. Antimatter does actually exist; it is not a figment of the imagination of science-fiction writers. All subatomic matter particles, such as electrons and protons, have their corresponding anti-particles. For example, the anti-particle of an electron is called a positron, which has the same mass as an electron but has the opposite electric charge. The other thing about matter and antimatter that is relevant to this discussion is that when they come together they go bang, inasmuch as they annihilate each other in a burst of pure energy. For example, when an electron and a positron come into contact with each other, their combined mass is converted completely into energy in the form of a burst of electromagnetic gamma radiation (see Chapter 6). In fact, this process of matter–antimatter annihilation releases more energy than any other reaction known to physicists. Thus the fusion reaction that powers the Sun is not the last word in energy generation; fusion converts only 0.7% of mass into energy, whereas the matter–antimatter reaction gives 100%! Thus if antimatter could be harnessed to power a starship, then maybe we can begin to compete with the Hollywood space engineers!

Before we get carried away, however, there are one or two issues about using antimatter that mean that such a propulsion system is a prospect only for the long-term future. The first of these is containment of the antimatter fuel within a matter starship. We can perhaps imagine a magnetic containment system, similar to that we have discussed for the fusion reactor, to prevent contact between the antimatter and matter parts of the ship. But what if the confinement process becomes unstable, which seems to be a fairly frequent occurrence on the starship *Enterprise*? The consequences for the ship and crew (and indeed for any nearby planet!) in this circumstance would be totally catastrophic. The other problem is that there appears to be not much antimatter around, which is fortunate. Small quantities can be obtained from experiments in terrestrial physics laboratories (such as the European Council for Nuclear Research [CERN] high-energy particle accelerator buried beneath the city of Geneva, Switzerland), but currently we have no means of industrial-scale production that would be required to produce anti-rocket fuel. As a consequence, we could say that antimatter is the most expensive substance on Earth.

Most of the starship rocket technologies that could be envisaged being developed this century (nuclear impulse engines and fusion-powered rockets, but perhaps not antimatter rockets) give speeds in the region of 10% of light speed—around 30,000 km/sec. Although this sounds fast, this can be considered to be something of a snail's pace given the scale of the universe. For example, if we were to send a starship to explore one of the approximately 2000 stars within 50 light years of Earth, the journey would take 500 years at this speed. One way to do this is to launch a large, self-contained interstellar ark, in which the ship would need to sustain many generations of crew, until finally the distant successors of the original crew finally arrive at the destination. This type of starship has become popularly known as a *generation ship*. However, one disadvantage of this type of ship is that it could easily be overtaken literally, and in terms of technology by a future starship with more sophisticated propulsion technology that left Earth at a later time. The crew of the generation ship would be shocked to find that their destination planet had already been colonized for many years by visitors from Earth! Let's take a brief look at some of these more exotic ideas for interstellar transportation.

Exotic Systems

It's interesting that many of the more exotic ideas for achieving interstellar travel derive their inspiration from science fiction. Perhaps the most obvious example of this is the warp drive, with which we have become familiar through Gene Roddenberry's epic *Star Trek* series.

Warp Drive

Miguel Alcubierre, a Mexican theoretical physicist, was intrigued by this idea, and in the early 1990s he set about attempting to find a way of describing how it might work within the formal framework of Einstein's theory of general relativity. As a consequence, he published a rather mathematical paper in 1994 in the learned journal *Classical and Quantum Gravity*, and since then his name has become synonymous with the idea of warp drive. As we mentioned in Chapter 1, Einstein's general relativity is essentially a new way of looking at how gravity works, which involves the notion that space and time are curved, or warped, by the presence of mass. Since Einstein also said that mass and energy are simply different forms of the same thing (see Chapter 6), space-time is also warped by the presence of energy. In Chapter 1 we saw how Newton's theory of gravity was overthrown by Einstein's new vision, in which the planets moved along their orbits around the Sun, like racing cars on a banked circuit, governed by the curvature of space-time produced by the Sun.

The key issue is how a warp drive–powered starship can travel at arbitrarily large speeds, effectively in excess of light speed, without violating the light speed limit. This sounds like an impossible trick, but there are ways to do this. The explanation resides in the more precise statement that, in general relativity, nothing can travel *locally* faster than the speed of light. In his deliberations about warp drive, Alcubierre found the illustration of the motion of objects in an expanding universe helpful in explaining his idea. So let's have a look at it to see if it helps.

One of the great triumphs of Einstein's theory of general relativity is that it predicts the expansion of the universe. This is one of the most profound achievements of theoretical physics in the 20th century, but it is also associated with an affair that Einstein himself considered to be one of his biggest blunders. Soon after the publication of his general theory in 1916, Einstein set about applying it to the universe as a whole, and showed that the universe would naturally be in a state of expansion. However, on the basis of the limited astronomical data then available, he was convinced that in fact the universe was actually static, so he introduced a new term into his equations of general relativity, essential a fudge factor, which he called the *cosmological constant*. With this new term in the equations he could model a static universe, in accord with what he then believed to be the case. However, in 1929, an astronomer named Edwin Hubble (after whom the famous space telescope is now named) published detailed observational evidence that, on the large scale, the universe is most definitely in a state of expansion. If only Einstein had believed his own analysis, he could have made one of the most profound predictions of theoretical physics, but it was not to be, and

Einstein had to return to his original equations, abandoning his cosmological constant (in fact, Einstein's cosmological constant has had a checkered history, and recent developments in theoretical physics have provoked scientists to consider its reintroduction into Einstein's theory).

What does it mean to say that the universe is expanding? A common misunderstanding is that the universe is effectively an infinite expanse of space-time, and at some moment in time, and at a particular point in space, the Big Bang happened. Thereafter, all the matter (galaxies) in the universe would appear to be moving away from each other and from a common point in space-time (the location of the Big Bang), gradually filling the huge expanse of space-time. However, the currently accepted view of the universe's expansion is subtly different from this. Rather than thinking of the universe as an explosion in a huge, fixed expanse of space-time, we have to envisage the universe—and by this I mean the fabric of space-time itself—as expanding. A good model of this, which is often quoted, is that of blowing up a balloon, although we have to lose a couple of dimensions. The four dimensions of space-time are now represented by the two dimensions of the rubber membrane of the balloon's surface. It is quite instructive to perform this simple experiment yourself, which emulates the expansion of the Universe. As we inflate the balloon, the rubber membrane (space-time) expands, and it is easy to see that each galaxy moves away from every other galaxy, representing Hubble's profound observational result that the balloon (the universe) is expanding.

Getting back to our attempt to understand Alcubierre's warp drive, it's important that you grasp the subtleties of the last paragraph, so if you haven't, I suggest you go back and give it another go. Because here's the point: as the balloon expands, "galaxies" on opposite sides of the balloon can move away from each other at speed, and yet at the same time they are stationary with respect to the rubber membrane (space-time). If we now contemplate the real, expanding universe in which we live, and think about the implications of this statement, we can have a situation where two galaxies are so far apart that the expansion of space-time itself causes their speed relative to each other to be in excess of the speed of light, while at the same time neither galaxy is *locally* exceeding light speed.

It is this kind of thinking that underlies Alcubierre's idea of how a warp drive–powered starship might work. He envisaged a drive system that warps the space-time surrounding the starship, in such a way that space-time is expanded behind the starship, and contracted ahead of it, as illustrated in Figure 11.2. The expansion of space-time behind the starship effectively pushes the departure point many light-years back, while the contraction in front of the vehicle acts to bring the destination similarly closer. The starship

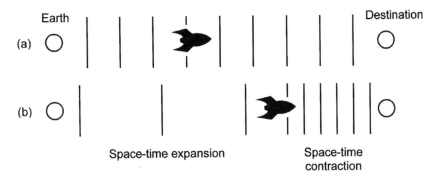

Figure 11.2: (a) Warp drive off: starship journeys through flat space-time. (b) Warp drive on: the space-time surrounding the starship is distorted to achieve faster than light speed travel.

itself is left in a locally flat region of space-time between the two warped regions. In this way, motion faster than light is possible, as seen by an observer outside the region disturbed by the warp drive, while at the same time light speed is not exceeded locally by the starship—a really neat idea!

Is warp drive technology a feasible means of interstellar travel? To warp space-time behind and ahead of the vehicle, we know that the drive system must be able to manage and manipulate huge amounts of mass and energy. (*Star Trek* fans now have some idea what the famous dylithium crystals do in energizing *Enterprise*'s warp drive.) Alcubierre wrote down the equations defining the necessary warp field for a starship, and then went on to investigate the kind of mass-energy source that would be needed to generate such a field. And here's the really bad news: the required space-time curvature needs the presence of a negative energy density. What this means is that the Alcubierre warp drive can be fueled only by a form of material that the scientists call *exotic matter*. Effectively, this comprises material that possesses characteristics such as negative mass, and there is debate among the experts about whether such matter even exists. Classical physics says it does not, whereas quantum theory says maybe it does. Either way, so far it is something that has escaped detection by scientists. Obviously, this is a bit of a blow to the feasibility of the Alcubierre warp drive, but his paper is probably not the last word on this topic. Fundamentally, we know that warp drive will be a difficult nut to crack, simply because of the huge amounts of mass-energy that is required to manipulate the curvature of space-time. However, I am sure alternate views will be presented in the future by theoretical physicists, and perhaps one of them will find a viable technical foundation for such a form of interstellar travel.

Wormholes

In many ways, wormholes are an even more curious idea than warp drive, but one that is a little easier to explain. Again, it is a technique that has its foundations in Einstein's theory of gravity for achieving super-light-speed travel without actually exceeding the light speed limit. Like warp drive, the wormhole concept has been grasped enthusiastically by science-fiction writers to overcome that awkward problem of how their space-faring heroes can travel with ease across the Galaxy. One such example, among many, can be found in Carl Sagan's novel *Contact*, in which the heroine Ellie travels 25 light years to the star Vega in the blink of an eye through a network of wormholes engineered by a long-lost civilization.

Space-time is not just a means of measuring where and when an event takes place, but it is also a dynamic entity that can be warped and curved by the presence of mass and energy. It has been known for many years that the equations of Einstein's theory allow solutions that permit space-time to be *multiply connected*. In other words, it allows for the existence of what are essentially short cuts through space and time, so that two distant regions of the universe can be connected by a much shorter higher-dimensional route. The term *wormhole*, first coined by physicist John Wheeler in 1957, has been universally adopted to label this curious feature of relativity theory, having its origins in the analogy used to explain the phenomenon. The usual analogy is to imagine a worm moving on the surface of an apple, starting out at a point A and moving to a point B that is on the other side of the apple (Fig. 11.3a). It has two choices: either it can go the long way round on the two-dimensional surface of the apple (route 1), or it can take a shorter journey (route 2) via a wormhole through the three-dimensional interior of the apple.

We can relate this analogy to a starship journeying between two points in space-time that are light years apart (Fig. 11.3b). It too can take the long route through space-time (represented by the two-dimensional curved surface) or use a conveniently located wormhole (represented by the three-dimensional passageway through hyperspace) to take a short cut. The wormhole allows the starship to effectively cover the distance at super-light speed, but without actually exceeding Einstein's speed limit. The analogy of the apple is helpful, but again we have lost a couple of dimensions in the discussion of the process.

So, is this form of interstellar travel a viable proposition for the future? Well, things are not quite as simple as the above analogy implies. One of the earliest wormhole solutions to the equations of general relativity was found by Einstein himself, in collaboration with a colleague Nathan Rosen, in 1935. This was christened an *Einstein-Rosen bridge*, but it took a good few years more for the theoretical physicists to realize that this type of wormhole was

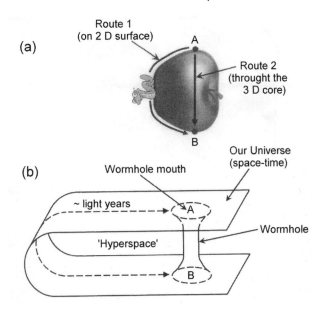

Figure 11.3: (a) Moving from point A to point B, the worm has a choice of taking the longer route on the apple's surface, or the shortcut through the middle. (b) This analogy is often used to illustrate the idea of a using a shortcut through a cosmic wormhole between two points A and B in the universe that may be light years apart.

unstable. It would close as soon as it was formed, making the transfer of people or starships through the cosmic passageway impossible. Since then, a great deal of work has been done investigating the stability of wormhole solutions of Einstein's theory, and again the bottom line is not good news for prospective interstellar travelers. The theory suggests that to keep a traversable wormhole open requires the use of exotic matter—the same stuff that we cannot find to power the warp drive! So although wormholes remain an intriguing prospect for the future, we seem to have hit the buffers again, with the engineering of such a scheme requiring huge amounts of negative mass and energy.

However, all is not lost. It should be borne in mind that the current wormhole solutions of Einstein's equations are based on his original classical theory, which focuses on the physics of the very large: planets, stars and galaxies, and the like. However, physicists are currently struggling to find a *theory of everything* that will describe the universe, not only on the large scale, but also on the very small scale where quantum mechanics presently reigns. Nobody really knows what such a theory will say about the future prospect of engineering a cosmic wormhole network as a kind of interstellar metro system!

Epilogue

Its time to wrap up this journey through spacecraft design. I hope readers have found it useful, and that it has fed their interest in space. For my part, I have found the process of writing enjoyable and quite therapeutic, a bit like a download of my interests, enthusiasms, and experience, and I hope in such a way as to make it accessible to people who do not have a technical background. I have found this aspect of trying to explain fairly complicated ideas in an informal and entertaining way challenging.

As I write these final few paragraphs, it is October 2007, which marks a significant anniversary. It is 50 years since the former Soviet Union lofted a small satellite called Sputnik 1 into Earth orbit, thus heralding the dawn of the Space Age! I recently read a quotation from Buzz Aldrin, the Apollo astronaut who followed Neil Armstrong onto the Moon's surface during the historic first landing in 1969. In 1957, Aldrin recalls, Sputnik 1 made no great impression on him: "It seemed little more than a stunt." It is easy to understand this reaction, considering that he was then flying fighters from bases in West Germany at the front line of the Cold War with the Soviet Union. No doubt the beep-beep signal from space seemed to him to be inconsequential compared to the reality of training for a conventional or even nuclear war in that region of central Europe.

However, this view wasn't shared by Buzz's bosses back in Washington, D.C. The Scientific Advisory Board Ad Hoc Committee on Space Technology met in December 1957 at the Department of the Air Force Headquarters in the aftermath of Sputnik. Their report (National Security Agency NSA 00600, dated December 6, 1957), once classified as secret but now released under the Freedom of Information Act, is summarized by the statement, "Sputnik and the Russian ICBM [intercontinental ballistic missile] capability have created a national emergency." To counter the perceived Russian threat, the committee recommended the urgent commencement of a number of active Air Force–led programs:

- A program to develop second-generation ICBMs having a certain and fast reaction to Russian attack
- The acceleration of the development of reconnaissance satellites
- The establishment of a vigorous space program, with an immediate goal of landings on the Moon

So the intention to land men on the Moon was on people's lips long before President Kennedy's famous speech of 1961.

The shock of a little piece of Russian technology over-flying U.S. territory, with the launch of Sputnik in 1957, rocked America, and the first manned flight of Yuri Gagarin in 1961 was like twisting the knife in the wound. How the story unfolded from there is well known, with the Cold War competition between the two superpowers driving the space race to the Moon, culminating in Armstrong's first lunar footprints in 1969. This all seems a distant memory now, but it is for me one of the most vivid, enduring, and inspirational memories that I have of the first half-century of the Space Age. However, as I have said before, the other overriding feeling I have is one of impatience, a feeling that we ought perhaps to have already sent astronauts to stand on more distant planets. It seemed that 15 years into the new Space Age, with the departure of Apollo 17 from the Moon's surface in 1972, manned exploration of space appeared to have almost stalled.

By comparison, the Aviation Age began in 1903 with a 30-mph flight of the Wright brothers' first heavier-than-air airplane at Kitty Hawk. From these humble beginnings, the development of aviation continued unabated, and if we take a snapshot of where things stood half a century later, the first jet-powered civil airliner was already in operation, and experimental aircraft had already flown at twice the speed of sound—fifty times faster than the Wright flyer, at around 1500 mph! If we could achieve a speed of fifty times faster than Sputnik 1 now, we could reach the orbit of Mars in less than 3 days!

Although this kind of argument is flawed, nevertheless it does point to the undeniable fact that the Space Age has differed from the Aviation Age. Looking forward to the next half century, I have a greater optimism that manned exploration of space will accelerate. With the restructuring of the American space program, brought about principally by the retirement of the space shuttle fleet in a few years and renewed international competition from nations such as China, I have confidence that we will return to the Moon and venture to Mars. The other big issue, of course, is the cost of reaching orbit, and again I feel optimistic that the problem of aircraft-like launcher operation to Earth orbit will be solved in the next couple of decades. I feel the time is right.

This is all a bit late for me and for my career in the space sector, but nevertheless I have greatly enjoyed the era of the boom in space applications—communications, navigation, and Earth observation—and space science. What we have learned about the universe through the eyes of instruments like the Hubble Space Telescope has been phenomenal.

I look to the future with optimism, and I hope that this book will play a small part in inspiring young people to get involved in space science and engineering.

Index